Organic Food, Farming and Culture

ALSO AVAILABLE FROM BLOOMSBURY

Food, Warren Belasco

The Food and Folklore Reader, Lucy M. Long

The Food History Reader, Ken Albala

Food Studies, Jeff Miller and Jonathan Deutsch

Organic Food, Farming and Culture

An Introduction

Edited By
JANET CHRZAN AND
JACQUELINE A. RICOTTA

BLOOMSBURY ACADEMIC
LONDON • NEW YORK • OXFORD • NEW DELHI • SYDNEY

BLOOMSBURY ACADEMIC
Bloomsbury Publishing Plc
50 Bedford Square, London, WC1B 3DP, UK
1385 Broadway, New York, NY 10018, USA

BLOOMSBURY, BLOOMSBURY ACADEMIC and the Diana logo are trademarks of
Bloomsbury Publishing Plc

First published in Great Britain 2019

A catalogue record for this book is available from the British Library.

Library of Congress Cataloging-in-Publication Data
Names: Chrzan, Janet, editor. | Ricotta, Jacqueline, editor.
Title: Organic food, farming and culture / edited by Janet Chrzan and Jacqueline Ricotta.
Description: London, UK ; New York, NY, USA : Bloomsbury Publishing, Plc, 2018. |
Includes bibliographical references and index.
Identifiers: LCCN 2018030755 | ISBN 9781350027848 (pb) | ISBN 9781350027831 (hb) |
ISBN 9781350055834 (epdf) | ISBN 9781350027855 (ebook)
Subjects: LCSH: Organic farming. | Natural foods.
Classification: LCC S605.5 .O679 2018 | DDC 631.5/84–dc23
LC record available at https://lccn.loc.gov/2018030755

ISBN: HB: 978-1-3500-2783-1
PB: 978-1-3500-2784-8
ePDF: 978-1-3500-5583-4
eBook: 978-1-3500-2785-5

Typeset by Newgen KnowledgeWorks Pvt. Ltd., Chennai, India
Printed and bound in the United States of America

To find out more about our authors and books visit www.bloomsbury.com
and sign up for our newsletters.

Contents

Illustrations

Figures

Tables

Contributors

E. N. "Gene" Anderson is Emeritus Professor of Anthropology at the University of California, Riverside. He received his Ph.D. in anthropology from the University of California, Berkeley, in 1967. He has done research on ethnobiology, cultural ecology, political ecology, and medical anthropology in several areas, including Hong Kong, British Columbia, California, and the Yucatan Peninsula of Mexico. His books include *The Food of China* (1988), *Ecologies of the Heart* (1996), *Political Ecology of a Yucatec Maya Community* (2005), and *The Pursuit of Ecotopia* (2010).

Jim Bingen is Professor Emeritus of Community, Food, and Agriculture in the Department of Community Sustainability at Michigan State University. He continues to work on several applied research studies of organic farming and place-named foods in the US states linked to Great Lakes and in Europe. He was a Fulbright Distinguished Chair at the University of Natural Resources and Applied Life Sciences in Vienna, and he was awarded the *Chevalier d'Ordre du Mérite Agricole* (Order of Agricultural Merit) by the Government of France.

Barrett P. Brenton, Ph.D., is Professor of Anthropology and Director of the Center for Global Development and Graduate Program in Global Development and Social Justice at St. John's University in Queens, New York. As a specialist in the nutritional anthropology of sustainable food systems, Dr. Brenton's extensive publication record and cross-cultural applied fieldwork experience across five continents ranges from studies of dietary biodiversity, food insecurity, and health in Indigenous Native American and Ecuadorian Amazon communities; to research on food aid, GMOs, and food sovereignty in Zambia; to community-based urban food systems and health disparities research in the New York City.

Leigh Bush earned her Ph.D. from Indiana University's (IU's) Food Studies program where she researched the effects of new media on the culinary industry. She has participated in wine, dairy, and meat production in Europe and the United States before doing her ethnographic research while working in the restaurant industry and at food media startups in Chicago. Leigh continued to explore food and media networks as an instructor at IU, as a fellow at the IU Food Institute and at the travel and exploration digital media company, Atlas Obscura/Gastro Obscura. She hosted the wine documentary *Hoosier Hospitality: Wine* and regularly guest-hosts WFIU's syndicated food radio program *Earth Eats*.

Christina Callicott holds an MA in Cultural Anthropology from the University of Florida (UF), with a forthcoming Ph.D. in the same. Her primary line of research involves the relationship between plants, music, and diet in the ethnomedical practices of Quechua-speaking peoples of the Peruvian and Ecuadorian Amazon. She has published on the ethnomusicology of Amazonian shamanism and the socio-environmental impacts of its internationalization. She is a long-time consumer of and advocate for local and organic foods and the decentralization of the food industry. She is a graduate of the UF Tropical Conservation and Development Program.

Janet Chrzan holds a Ph.D. in Physical/Nutritional Anthropology from the University of Pennsylvania, where she is Adjunct Assistant Professor of Nutrition in the School of Nursing. Her research explores the connections between social activities, nutritional intake, and health outcomes. She has published articles on food use, nutrition education, health and culinary tourism, and is the author of *Alcohol: Social Drinking in Cultural Context* (2013) and co-editor of the three-volume set *Research Methods for Anthropological Studies of Food and Nutrition* (2017). In 2007 she founded the Oakmont Farmers Market and the Haverford Township Farmers Market Association, a community not-for-profit organization that supports sustainably farmed local food and community health education. She is past president of the Society for the Anthropology of Food and Nutrition and has served on the boards of several health, agriculture, and nutrition-related academic and community organizations.

Budd Cohen graduated from Baltimore's International Culinary College in 1990 to work as the executive chef at the Commissary Restaurant and then opened several restaurants for Neiman Marcus; he was the executive chef for the Sheraton Plaza at Valley Forge and joined Williamson Hospitality in 1998, which was bought by Culnart Inc. He has dedicated his culinary career to promoting the connection between local farms and consumers. His focus has been to provide nutritious local farm-direct foods to educational settings ranging from pre-kindergarden to medical school; his ability to build relationships with farmers, consumers, producers, and distributors has helped change the culinary landscape in schools. He enjoys sharing his passion with all interested parties and offers an alternative path to the culinary career by supporting the local farm community. He is the author of *Farm to Table for Schools*.

Kathleen Delate's current position as professor at Iowa State University is a joint position between the Departments of Horticulture and Agronomy, where she is responsible for research, extension, and teaching in organic agriculture. She was awarded the first faculty position in Organic Agriculture at a Land Grant University in 1997. She has a BS in Agronomy and an MS in Horticulture from the University of Florida, and a Ph.D. in Agricultural Ecology from the University of California, Berkeley. She has farmed organically in Iowa, California, Florida, and Hawaii. She spent a sabbatical leave in Italy in 2014, studying organic farming, with some of the 48,000 organic farmers there.

Jonathan Deutsch, Ph.D., is Professor of Culinary Arts and Food Science at Drexel University. He oversees the Drexel Food Lab, a product development and food innovation lab that solves real-world problems in sustainability, health promotion, and food access. He is the author or editor of eight books, including *Barbecue: A Global History* (with Megan Elias), *Culinary Improvisation*, and *Gastropolis: Food and Culture in New York City* (with Annie Hauck-Lawson). When not in the kitchen, he can be found behind his tuba.

Adam Diamond is an environmental and food systems consultant. He does policy analysis, research, and proposal development pertaining to agricultural policy, organic farming, regional food systems, and composting. He has taught at American University and George Washington University on environmental politics, the political economy of agriculture, and agricultural policy. For six years Diamond worked at the USDA's Agricultural Marketing Service, where he did applied research on local and regional food system development. Diamond received his Ph.D. in geography from Rutgers University in 2006.

Paul Durrenbereger, an anthropologist, has done fieldwork in highland and lowland Thailand, Iceland, Iowa, Alabama, Mississippi, Pennsylvania, and Chicago. He has taught for twenty-five years at the University of Iowa and fifteen years at Penn State before retiring to live with his wife, Suzan Erem, in rural Iowa. He has written numerous papers in popular anthropology journals and a number of books, and served as president and board member of a number of academic organizations. He now serves on the board of directors of the Sustainable Iowa Land Trust.

Steven Eckerd is Executive Chef of the Pub and Kitchen restaurant group, which operates three restaurants in the Philadelphia region. Born in rural Etters, Pennsylvania, Eckerd grew up with a fascination for farming and began his culinary journey young, educating himself on how things grow and exploring ways to prepare the garden's harvests as the seasons changed. After completing his education at the Culinary Institute of America in Hyde Park, New York, Eckerd joined the team of world-renowned Chef Daniel Boulud at NYC's Daniel. Eckerd first came to Philadelphia in 2010 and has worked in some of the city's finest restaurants, including Vetri, Osteria, Le Bec Fin, Little Fish, and Lacroix at The Rittenhouse. His culinary focus remains tied to the farm-to-table movement, convinced that local ingredients provide true inspiration to a chef.

Suzan Erem has worked as a union organizer and communicator in Iowa and Chicago and as a journalist. She wrote *Labor Pains: Inside Americas New Union Movement* (2001) and *Do I Want to Be a Mom: A Woman's Decision of a Lifetime* (2003) and has co-authored "On the Global Waterfront: The Struggle to Free the Charleston 5" in *Monthly Review* (with Paul Durrenberger, 2008) and many books and journal articles. For seven years, she was editor of the alternative newspaper *Voices of Central Pennsylvania*. She is the founder and president of the board of directors of the Sustainable Iowa Land Trust. She lives with her husband, Paul Durrenberger, in rural Iowa.

Valentin Fiala completed his masters at the University of Natural Resources and Life Sciences in Vienna (BOKU). After working for four years on rural regional development near his hometown just northwest of Vienna, he is currently a Ph.D. candidate at the Division of Organic Farming at BOKU and also a research assistant of the working group Transdisciplinary Systems Research, a part of the division. In his research he deals with the institutional development of organic farming in Europe, the application of system theories and methods, as well as societal images and media representations of organic farming.

Bernhard Freyer has been the Head of Division of Organic Farming, University of Natural Resources and Life Sciences (BOKU), Austria, since 1998. He is also head of the working group Transdisciplinary Systems Research and a senior fellow at University of Minnesota, Minnesota Institute for Sustainable Agriculture. His research topics are systems theory, ethics, transdisciplinary research, environmental sociology, organic farming and societal discourse, farming and food systems research (temperate, tropical, and subtropical regions), and cropping systems.

Preety Gadhoke, Ph.D., MPH, is Assistant Professor of Global Health and teaches public health and health systems courses for the Master of Public Health program and undergraduate allied health programs at St. John's University, New York. As a public health scholar, Gadhoke engages in theoretically grounded and innovative methodological research to inform culturally sensitive and relevant global health and nutrition interventions for marginalized populations. Her current research includes youth access to mental health services among Shinnecock Indians, community-based health and nutrition among Indigenous Shuar communities in Ecuadorian Amazon, child as change agent approach for adult obesity prevention in Native American households, and food systems, food security, and health disparities in New York City. Past research includes artisanal small-scale gold mining and health in Mozambique and social network analysis for vaccine introduction in Nigeria. Prior to joining the academy, Gadhoke worked at the U.S. Government Accountability Office, Pan American Health Organization, Centers for Disease Control and Prevention, county health departments, and in the pharmaceutical industry.

Mette Weinrich Hansen is an Associate Professor of Sustainable Food Networks. Her research and teaching work is interdisciplinary, involving socio-technical aspects of food production. In several research projects and in her Ph.D. work she has been using qualitative methods in studying alternative food networks and value aspects of organic food production. Since 2009 she has been involved in the organization and teaching of the Integrated Food Studies program at Aalborg University Copenhagen. She is affiliated with Copenhagen University, Department of Food and Resource Economics, section for Consumption, Bioethics and Governance as an associate professor where the educational program and the research has moved. Currently she is working in the areas of public food systems and sustainability and innovation in small- and medium-size food enterprises.

Joseph R. Heckman, Ph.D., is Professor of Soil Science, Rutgers University, and teaches courses on soil fertility, organic crop production, and agroecology. He conducts research and extension programs on optimizing nutrition and soil quality in support of plant, animal, and human health. He has served as chair of several professional organizations, including Council on History, Philosophy, and Sociology of Soil Science; Committee on Organic and Sustainable Agriculture; Organic Management Systems Community of the American Society Agronomy; and as president of the northeast branch of the Crops, Soils, and Agronomy Tri-Societies. Heckman has authored numerous publications on soil fertility and organic farming.

Mark Keating began working in organic agriculture as a farmworker in 1988 and continues to spend as much time as possible with farmers in the field. He has also worked as a cooperative extension agent, organic inspector, university lecturer, policy advocate, and journalist. He was the lead organic crop-and-livestock specialist with the USDA National Organic Program between 1999 and 2002 during its implementation. He subsequently served with the USDA Marketing Services Branch through 2004 to develop farmers' markets and other direct-to-consumer sales initiatives. He currently works nationally as a consultant on organic, local, and sustainable food initiatives.

Milena Klimek is a doctoral candidate of the Division of Organic Farming at the University of Natural Resources and Life Sciences, Vienna, where she completed her masters and teaches a course on ethics in sustainable agriculture. As an Austrian American she finished her undergraduate degree at St. Olaf College in Minnesota. Her project work includes food and farming issues around Europe, the United States, and developing organic curriculum in Armenia. Her main research interests focus on the values and beliefs of people related to nature, food, and farming; agroecology; farmer/consumer patnerships; the future of farming; and transdisciplinary qualitative research methods.

Niels Heine Kristensen (NHK) is Full Professor in Culinary Arts and Food Innovation at Umeå University, Sweden. Until 2017 he was Professor in Food Policy and Innovation at Aalborg University Copenhagen, Denmark, where he in 2009 initiated the Food Studies program, which included the Master in Integrated Food Studies and the Research Group Foodscapes, Innovation and Networks. From 1999 to 2009 Kristensen was Associate Professor at the Department of Management Engineering–Innovation and Sustainability, at Technical University of Denmark (DTU). He earned his Ph.D. from the Department of Social Sciences at the DTU on sustainable food systems involving socio-technical aspects of food production. He has published on food culture, public food systems, alternative food networks, and organic and sustainable transition in the agro-food sector, and he has served as an expert, consultant, and reviewer on a number of councils, ministries, and boards, including the European Commission, the Ministry of Agriculture, Food, and Fisheries; the Ministry of Environment; the Ministry of Industry; the Danish Board of Technology; nongovernmental organizations, regional, and local bodies, and companies. Kristensen is currently involved in a number of Horizon2020 research networks and projects such as GLAMUR and NEXTFOOD.

John T. Lang is an associate professor of Sociology at Occidental College in Los Angeles, California. He is interested in the study of food as a lens for investigating questions that lie at the intersection of multiple areas such as consumption, culture, risk, trust, and the environment. Lang's work has been published in venues including *Food Policy, Gastronomica, the International Journal of Public Opinion Research, Risk Analysis,* and *AgBioForum*. His book, *What's So Controversial about Genetically Modified Food?,* explores the science—and myth—that surrounds genetically modified food in order to help us understand just what is at stake.

Jacqueline A. Ricotta, Ph.D., is a professor of horticulture at Delaware Valley University in Doylestown, Pennsylvania. She teaches courses in botany, sustainable agriculture, organic food and fiber, integrated pest management, commercial vegetable production, and marketing of horticultural products, and was involved in bringing certified organic to the university's farm in 2004. She helped create the Organic Farming Certificate Program (in partnership with the Rodale Institute) as well as the Sustainable Agriculture Systems major and the Food Systems minor. In 2012, Ricotta received the DelVal distinguished faculty member award.

Richard Robbins received his Ph.D. in anthropology from the University of North Carolina and has spent his entire teaching career at SUNY, Plattsburgh. His recent publications include *An Anthropology of Money: A Critical Analysis* (with Tim DiMuzio, 2017) and *Debt as Power* (also with Tim DiMuzio, 2016). Other works include *Cultural Anthropology: A Problem-Based Approach* (7th edn., 2016) and *Global Problems and the Culture of Capitalism* (6th edn., 2014). He is the recipient of the American Anthropological Association/Oxford University Press Teacher of the Year Award. He is currently SUNY Distinguished Teaching Professor at Plattsburgh and teaches a course on the anthropology of food.

Erika Tapp has worked in community-based nonprofits in North Philadelphia for the past ten years. These organizations have focused on addressing the needs of low-income communities with a particular focus on development and environmental issues in urban environments. Presently, she is teaching in the MBA program at Villanova University as part of their Social Enterprise curriculum. Erika earned her Bachelors of Architecture from Cornell University, a Masters of Arts in Architectural History from the University of Pennsylvania, and her Masters of Business Administration from

Villanova University. Her spare time is spent trying new restaurants and cuisines, as well as tending her own garden.

Catherine M. Tucker is an ecological and economic anthropologist at the University of Florida. Her research focuses on community-based natural resource management, food systems, and environmental governance. She is currently working with Central American coffee producers to explore how alternative trade (including organic certifications) affects their livelihoods, environmental sustainability, and adaptations to market volatility and climate change. She is the author of *Coffee Culture: Local Experiences, Global Connections* and *Changing Forests: Collective Action, Common Property and Coffee in Honduras*. Her recent work has been published in *Human Ecology, Human Organization, Global Environmental Change, Ecology and Society, Environmental Science and Policy*, and *Society and Natural Resources*.

Bob Turnbull's current position is a program coordinator for Iowa State University's Organic Program. In this capacity, his duties include designing and conducting field research, grant writing, literature review, and coordinating outreach programs aimed at fostering understanding of organic agricultural production, marketing, and law. He graduated from the University of Florida with a BSBA and from Golden Gate University with a JD.

Alex Wenger is the founder of the Field's Edge Research located in Lancaster County, Pennsylvania. He is engaged in plant breeding and seed trials, with the goal of adapting varieties for sustainable farming systems. He is involved in preserving and promoting forgotten foods, alongside "old-time" mentors, research scientists, and farming advocates. He also supplies specialty vegetables and culinary crops to high-profile chefs in Lancaster, Philadelphia, and New York City. Presently, he is completing his MS in Sustainable Food Systems from Green Mountain College and has plans to pursue a Ph.D. focused on plant breeding and ethnobotany.

Foreword

The story of organic food started a long time ago, in fact way back in 1942, when J. I. Rodale coined the words "organic agriculture" as a way of differentiating his thoughts on food production from those of most agricultural professionals. He thought we should focus our attention on the soil and not the inputs if we were to achieve our goals of producing a healthy population. My journey to organic farming started over forty years ago, when I first began my career as both a young farmer and, at the time, a field foreman for the prestigious Rodale Institute. It became evident to me, even back then, that the path that has led us to our current conventional food production system may have been the wrong path. Jump ahead to today and I'm the Executive Director of Rodale Institute, and along the way I've been involved with organic standards working with the National Organic Standards Board, with certification through the creation of Pennsylvania Certified Organic, with research developing organic no-till systems. And, I've had the pleasure of working with some of the brightest, hardest-working and most committed individuals anyone could imagine—especially people like Jacqueline "Jackie" Ricotta and Janet Chrzan, the editors of this book.

These are people who started their career much differently than I did; however, they began to see that the path we are on was not the path that would lead us to where we thought we should go. And, they began to ask questions. This book is all about questions—questions we need to ask and questions we need to answer. This book is also written specifically to inspire students, teachers, researchers, farmers, and simply those who are curious to ask their own questions and then find the answers. Questions and answers of this magnitude take a combined effort; so this book also builds on another great strength of the editors—the strength of bringing together and unifying unique contributors. By bringing together a multitude of voices a true chorus has emerged—a chorus of voices that ties together the complex pieces of an organic food system. From systems standards to the consumer's plate, the reader will take a journey through time to look at the history of organics, a journey through the science of organic food production and nutrition, a journey through the sociology of rural communities and their impact on the transition to organic production, and a journey through marketing as the book explores the linkages between consumers and the soil.

True organic production has always been about feeding a growing world population by placing our energy on feeding the soil. By striving for healthy soil, we inevitably reach our goal of healthy people and a healthier planet. This seemingly simple concept isn't achieved without meeting and overcoming serious challenges. And, this book doesn't shy away from identifying and addressing these challenges. The idea of linking a philosophy with systems of production isn't new, but it can be difficult to garner wide acceptance of the relationship. It can be challenging to gain support for the idea that we may sacrifice short-term yields for long-term resiliency in the system. Or that we need to gain consumer trust in voluntary regulations. And finally, readers of this book will be encouraged to think beyond organic to a system that includes the issues of social justice and equitable food distribution.

As a society, we all need to begin planning for the future we want. After all, we'll get the food system we work for. A food system that has no pesticide residues, a system that supplies nutritious food with a complex chemistry only the soil can supply, a food system where consumers know the farmers who grow their food, and a food system that places value on protecting or enhancing the resources it needs to function—all are within our reach.

How we gain access to the products of this new food system will also be important. Millennials are the New Boss and farmers, marketers, and policy-makers need to refocus on the values they bring to the conversation. Today more than ever it truly is "all about the label." But what's behind the label is what's important. Readers will explore the challenges of educating consumers about this very point—what does the label really mean versus what does the marketer want consumers to believe the label means?

Through *Organic Food Farming and Culture*, Janet Chrzan and Jacqueline A. Ricotta have taken a giant step in helping us all understand how we got to this point in our journey and now have inspired us all to refocus our energies to create a new path. It is my hope that each reader feels the need to ask their own questions, to discover their own answers, and to add to the fabric of organic food in our culture.

Jeff Moyer
Executive Director of the Rodale Institute

Acknowledgments

The writing of this book has involved the collaborative efforts of many talented people. We are grateful for the enthusiastic response to requests to write chapters and narratives, and appreciate the patience of the staff at Bloomsbury as we collated and edited this volume.

The idea for this volume was suggested to Janet Chrzan in 2006 by Kathryn Earle, who was then Managing and New Business Director of Berg Publishing (now Bloomsbury). It was a great idea, and Janet said no immediately, knowing that she lacked the expertise to cover the subject appropriately. Later, under the tutelage of Louise Butler, commissioning editor, the proposal was developed to include a wide variety of voices and topics. Additional wise guidance was provided by commissioning editor par excellence Jennifer Schmidt, and the proposal was largely written by Richard Robbins. Realizing that agronomics truly was essential, Jackie Ricotta was asked to join as co-editor, thus providing the much-needed expertise in organic farming and agronomics. Editor Miriam Cantwell came on board at the end and has displayed immense patience with the project's occasionally faltering steps. Without the guidance and enthusiastic mentoring of every one of these individuals, this volume would not have been possible, but a very special "thank you" is offered to Richard Robbins, who has been Janet's editor, mentor, and enthusiastic champion throughout this project and others.

The authors in this volume are accomplished scholars in their respective fields, and they have been willing accomplices in the creation of a project with a distinctly peculiar written structure. We thank them for their exacting and inspiring research and their patience as this project stretched past original due dates. We must especially thank E. N. "Gene" Anderson, whose promptly submitted and superb chapter created the intellectual path forward and proved that the volume could truly come into existence. Additional thanks are due to Jeff Moyer, Executive Director of the Rodale Institute, for inspiring farmers everywhere with Rodale's agronomic studies and graciously providing this volume's Foreword. Deep thanks are due to Leigh Bush, Steven Eckerd, Erika Tapp, and Alex Wenger, who provided the four narratives designed to link the academic chapters to the real world of eaters, thinkers, farmers, and chefs. They provided brilliant and inspired true-life narratives and analyses, schooling us in what was important about each chapter and giving their time and stories with intelligence and grace.

We must also thank the producers and patrons of the Oakmont Farmers Market, who have collectively provided hundreds of hours of conversation and interviews about what it means to be a customer, a farmer, and a food activist. Special thanks to Lisa and Ike Kirschner of North Star Orchard, who have guided Janet intellectually in her attempt to understand how local and organic farmers grow a successful business and whose farm brilliantly supports the ideals of the "Triple Bottom Line": People, Planet, Profit. In addition, thanks are due to William Woys

Weaver for introducing Janet to Alex and Steven, and for countless conversations that shaped her understanding of food and farming in Pennsylvania.

Jackie is eternally grateful to the many organic and sustainable farmers of Bucks County, Pennsylvania, who have welcomed her on field trips to their farms—they have provided inspiration to hundreds of students over the years, and Jackie never fails to learn from them. Past and present board members of the Bucks County Foodshed Alliance have been stalwart proponents of organics and friends whose advice and tireless support of local agriculture have been a joy to see. The partnership between DelVal and the Rodale Institute continues to strengthen and grow education in organic agriculture, as well as train new organic farmers, and Jackie is grateful for the support and encouragement of both organizations.

We would like to extend our deepest appreciation and gratitude to our families, colleagues, and friends who fortified our resolve to complete this project and provided support and encouragement. The ultimate inspiration for this tome was from the many organic farmers, toiling daily to provide our sustenance and nourishment while respecting the environment and leaving our world a better place. Our respect and admiration are here in these pages, and our sincerest hope is for their continued success.

Organic Food, Farming and Culture: Introduction

Janet Chrzan and Jacqueline A. Ricotta

Imagine, for a moment, that you are in a grocery store, a "big box" store, or food co-op, standing in the produce section. You are surrounded by heaps of apples and oranges, strawberries in clear plastic boxes, lettuces moist from misting, piles of lemons and limes, and watermelon sliced and whole. Every color and every shape of fruit and vegetable is enticingly arrayed in front of you, a dramatic display of options ... and you must choose what to eat. On one side of the section is a smaller area labeled "Organic Fruits and Vegetables"; you inspect what is offered. There is less variety, and less of each type. Some of the vegetables and fruits aren't as shiny as those on the other side of the vegetable section, and some seem to have a few blemishes. Still, you've heard that organic food is better for you, and better for the planet; so you think that maybe you'll buy organic this time. But then you look at the prices and realize that they are higher—sometimes a fair bit higher—than the other vegetables at the market; so you calculate again ... What to do? You defer this decision for a moment while you head over to the dairy section where, again, you are presented with this choice, since organic milk and dairy has become available in most grocery stores. Again, you notice a price difference, this time even more dramatic. It seems to be a pattern—the meat section also provides organic, and again, at a higher price than the other options. Perhaps you can't escape making this choice about your food, but the decision process is difficult. Do you buy the organic products that you suspect you "should" be buying, and maybe get less because of the price, or buy more of the "regular" dairy, meat, and produce, remembering that your food budget is, well, a budget and subject to constraints. You start thinking about what you've heard about organic foods, and why some of your friends say they only buy organic. You think about what you read online from the Environmental Working Group, the "Dirty Dozen," and the "Clean Fifteen" and start to worry about doing the right thing. You remember hearing an NPR interview about how pesticides are dangerous for farm workers, and worry even more, and not just about yourself. And then you remember reading Rachel Carson's book about pesticides, *Silent Spring*, and think about how waterways and animals are affected by what we spray on fields and

FIGURE 0.1 *Supermarket produce. Photo courtesy of Janet Chrzan.*

food crops. But still, you worry about the extra cost. Can you afford to buy organic? Can you afford not to?

Or perhaps you are visiting a farmer's market—maybe you are lucky enough to have a market in your neighborhood. You hear the customers around you talking excitingly about how everything is "local and organic" at the market. And yes, you agree, local and organic is the best, but you're not quite sure why; all you know is that everyone you know who's really into food prefers local and organic. But you look around you, and some of the vendors have "Certified Organic" written on their tent banners, but some do not. What does that mean? Aren't all farmers at a farmer's market supposed to be "local and organic"? Isn't that what everyone assumes? But what if they are not organic, should you buy just because they are local? Or should you always buy organic, even if it's not local?

This volume is not designed to provide answers to such difficult questions, but to provide you with authoritative and reliable background information about how to think through these issues so that you are better able to make decisions. There are so many variables to understand when choosing food these days, especially for those who live in systems of immense consumer abundance in the Global North. Perhaps more than any other time in our evolutionary history, human beings have access to more kinds of food throughout the seasons than ever before. Global supply chains make strawberries available to Wisconsin in January and autumn-picked

FIGURE 0.2 *Supermarket organic potato chips. Photo courtesy of Janet Chrzan.*

apples can be purchased year-round because of controlled atmosphere storage. Our diet has changed immensely because of globalization and technology, adding layers of choices and layers of decisions to be made. After all, we eat every day, and choosing what to eat involves thinking though any number of variables about our food, which include what we like to eat (which is the easiest to decide), nutritional value, cultural value (does it conform to our culture's concepts of "good to eat"?), where it came from, how it was produced, who produced it and how they were treated, how it was marketed, shipped, prepared, and, finally, placed on our plate. For many consumers, an important characteristic of how it was produced is increasingly whether it is certified organic or grown using organic standards.

Buying organic is a choice being made by more and more people each year. The US Department of Agriculture (USDA) recently released 2016 data about the growth of organics: 24,650 certified organic operations in the United States, and 37,032 around the world, a 13 percent increase from 2015 and about double what it was ten years ago. According to the Organic Trade Association, consumers of organic products have spent $43 billion in 2015; organic businesses grew by 12 percent and exports by 60 percent. Organic products can be found just about anywhere, from farm markets to big-box retailers and grocers to convenience stores. There is organic clothing, bedding, health and beauty products, mattresses, and even organic cigarettes. From a fringe movement in the 1960s to the mainstream choices of food and fiber today, organics has arrived.

While these numbers are impressive, the impact of organics goes much further. Organic farming has changed the way farmers of all types think about, work with, and protect their soil. No-till farming and the use of cover-crops is growing among all crop producers, and soil as a living and precious resource is finally gaining the respect that the organic pioneers Sir Albert Howard and Jerome Rodale wrote about over fifty years ago. The crop protection marketplace for both organic and conventional farms now includes dozens of new biocontrol products, a nod to consumer and workplace demands for safer, more wholesome foods. Organic farming has legitimized the concept of working with nature rather than trying to control it. And finally, as we face an unknown future concerning the impact of human activities on the fragile environment, organic farming is a means of sequestering carbon in the soil and slowing down the presently blistering pace of greenhouse gas emissions and climate change. As the Rodale Institute states, "There is hope right beneath our feet."

That hope extends to the multitudes of beginning farmers worldwide. Organics has provided a means for small, niche farms to have a role in the marketplace. No longer is it necessary to have hundreds of acres and a fleet of tractors. With the help of agencies such as the USDA and WWOOF (World Wide Opportunities on Organic Farms), organic farmer training programs and learning experiences have grown as people of all ages, races, and backgrounds embrace the opportunity to work with the land and produce food that is valued and revered by the eater. Everyday eaters as well as students in higher educational agricultural or food studies programs, now more than ever, need to learn both the conventional and organic perspectives of food production to be able to understand current research concerning food and agriculture. There is a large and growing number of such courses and programs in the United States and abroad, and immense student and faculty interest in using organics and sustainability as a lens with which to better understand the food system. This volume is meant to meet that need for students as well as eaters/consumers who are curious about how food is produced and what that means for them, for their families, their communities, and the planet.

This volume provides a unique and deliberately interdisciplinary approach to organic food and farming by including chapters written by agronomists, sociologists, nutritionists, and anthropologists as well as narratives provided by community members who are deeply invested in food and farming. Four food activists were tapped to contribute information about their experiences and perspectives about organic food and farming and to provide a deeper examination of the chapters. These activists include an organic farmer, a chef, a graduate student in food anthropology, and a community activist with a deep interest in growing, sourcing, and eating healthy food. We wanted to know what "organic" means to a farmer, a restauranteur, a food studies student, and to a food buyer. How do they sort through the thicket of information available about food? How do they know what they are eating is the "right food to eat"? Their reactions to the chapters include stories from their own experiences working with food and also what they thought were the most interesting messages to be found in each chapter. Their contributions demonstrate that the scientific and philosophical analyses presented are deeply relevant and interesting to everyday readers and eaters.

This volume includes topics as diverse as a history of the organic pioneers and its relationship to the environmental movement, to agronomic practice in developing and developed nations, to the marketing and adoption of organic food by consumers and institutions, to the impact of organic foods on health and food security. The deliberately broad range of subjects covered provides a

FIGURE 0.3 *At the Farmers Market. Photo courtesy of Janet Chrzan.*

balanced and empirically grounded introduction to organic food and organic food production. Because the topics are so broad they have been divided into four sections, each designed to highlight a particular cluster of behaviors linked to the practice of producing and using organic foods. The first section covers the deep, cross-cultural history of organic farming, its modern redevelopment in the twentieth century (what our authors label the "new organics"), and organic farming's links to the environmental movement in the United States and abroad. The second section focuses on practice—from the agronomic science that informs organic farming in both the Global North and Global South, to the value chains that determine the shape of the markets and consumer access, to organic products via distribution networks for farmers' markets and the dairy industry. The section concludes with a case study of two towns in Norway that have used civic structures to make organic food available in metropolitan institutional settings. The third section queries the values attached to organics, from the philosophical to the monetary. Starting from the bedrock, the cost, and value of land, this section further examines the perceived health value of organic food to consumers, the importance of organic and agroecological processes to farmers' economic success and adaptation to climate change, and concludes with a philosophical and anthropological analysis of how our culture values organic farming and food as both concept and identity practice. The final section posits organics as an element of the food system, both literal and metaphorical, querying how chefs and food buyers incorporate organic foods into institutional kitchens, and how the marketing and selling of organic food is constructed and maintained in capitalistic food chains. Readers are also challenged to ponder a part of the food system often conceptualized as opposite to organics—genetically modified organisms (GMOs). How does an understanding of organics inform what we think about GMOs, and how can learning about GMOs deepen our knowledge about organics? The volume concludes with a chapter that asks the reader to conceptualize the future of organics within a fully integrated food and farming system, as a method for growing food and fiber that embraces the best of traditional farming techniques while simultaneously incorporating new agroecological science. How can organic practice blend and further develop the old and the new to provide a path to greater sustainability, stronger community, and improved health for producers, consumers, and citizens around the world?

This book is designed for scholars of food studies, undergraduate students, those of us who eat every day and must make complex decisions about what foods to buy and why to buy them; for anyone concerned about or even interested in our food system, our farms, and the workers growing our daily meals. We hope that such an objective overview of organic food and farming will be of real benefit to anyone concerned with the food they purchase and consume, the environment, and how reenvisioning agricultural production practices can change the world ... and, of course, how to think through that thorny question "Organic or Conventional?"

PART ONE

Organic Farming: A History

EDITORS' INTRODUCTION

Whether delivered by truck, purchased in a local market or an obscure market across the world, or plucked fresh from a farmer's field, there is something universal about food. All cultures celebrate with food, and there are numerous dishes, traditions, and even fruit and vegetable varieties that hearken back to a time when farms and gardens were part of daily life rather than an anomaly to visit for entertainment. It is an interesting thing about organic food—approximately a hundred years ago, all crops were grown organically; we did not have synthetic fertilizers, pesticides, or GMOs. Animals were raised on pasture because that was what was available and most convenient. Fertilizer was synonymous with human or animal waste, not a manufactured product, interchangeable with explosives. But somewhere along the way, it became the norm for farms to be chemically intense and to use technology that perhaps had questionable safety for the environment, the growers, and the eaters. Growing organically became the darling of the countercultural revolution of the 1960s and never lost that persona of being on the outside—until now. This section tells the story of the genesis of organic food from the viewpoint of indigenous people in China, Southeast Asia, Mexico, and South America; of the organic pioneers, the environmentalists; as well as of farmers, eaters, and chefs. The food production systems that sustained populations for years have come full circle to where we now have—the need to certify and label our organic food—and leave unexamined and unmarked the food produced in large monocultures using synthetic pesticides, fertilizers, antibiotics, and genetically modified varieties. From learning about traditional, intensive organic farming to meeting the people who viewed these systems through the lens of increasingly regimented agriculture, to the social and political movements that shaped our times, this section chronicles the story of organics and the farming practices it is based on.

FIGURE I.1 *Spannocchia farmland. Photo courtesy of Jacqueline A. Ricotta.*

1

Profile: Alex Wenger, Organic Farmer

There is nothing as grounding as the moment that I plunge my hands into the earth and pull out glistening French Breakfast radishes, slice crispy leaves from sour sorrel plants, or plant sand-like mizuna seeds, watching them sprout through the soil and transform into beautiful vegetables within a few weeks. It is also extremely satisfying to share the fruits of my labor with others, watching chefs jump in circles because you have handed them the most stunning vegetables they've ever worked with. Farming is a beautiful way to live, but it is a lifestyle that can also be very painful.

It would be easy to date my first "farming" experience back to 2010, when I planted over an acre of exotic hot peppers, uncommon vegetables like papalo, bitter melons, and near-extinct heirloom varieties of maize, apples, and winter squash.

My agrarian background is rooted several generations in the past. I can remember my grandparents' stories of their farm. They grew fresh-cut flowers for their entire lives. Several acres were complemented by glass greenhouses, always awash in bright, beautiful blooms. My grandparents were specialty crop-growers. They chose to grow flowers on a small acreage when many of their neighbors in rural Lancaster County focused on mass-producing grains, dairy products, and vegetables. Specialty crops often require much more labor to produce, but sell for a higher price. The mass-produced commodity crops are easier to grow in large volume, but sell for significantly less. Often, commodity farms reach a scale of operation where they become ecologically unsustainable as well.

But in my grandparents' generation, no one talked about sustainability. Families passed land from generation to generation, and it was understood that the family as a whole had to take care of their patch of ground, so that they could pass along a successful livelihood to their children. Locally people used to call this "planting pears for heirs," because old-fashioned pear trees once took several decades to produce their first heavy crop. This is exactly what previous generations did for my grandfather. He continued that tradition by using cover-crops and building the health of his soil. Every year he would plant turnips in the fall, to return organic matter to the earth, and to eat as winter vegetables.

FIGURE 1.1 *Alex Wenger of the Fields Edge Farm. Photo courtesy of Janet Chrzan.*

FIGURE 1.2 *Alex Wenger and Steve Eckerd at the Fields Edge Farm. Photo courtesy of Janet Chrzan.*

When my grandparents sold their farm and moved to our current homestead, we felt the devastating consequences of neglected land. Our new property had been rented corn fields for over sixty years. Heavy chemical use and short crop rotations on what was always rented ground made it a no-man's-land. As we tried to farm, we clearly saw how damaging chemically driven food production could be. The goal of everyone who had used this land was to make a profit year after year. They were not invested in the long-term health of the soil. Years of weed jungles and concrete-hard ground left a soil where crops would not sprout, and stunted vegetables were now our challenge to manage.

Over the years, as my parents and I learned more about farming, from our grandparents and others in the organic community, the soil began to improve. Today I grow over eighty different crop varieties in any one year, supply multiple restaurants, work with rare seeds, and breed new food crop varieties. The niche market farming operation that I have built with the help of my parents was like that of my grandparents. What began as a series of home-school and undergraduate research projects has become "The Field's Edge Research Farm." My mission is to educate food professionals, consumers, and other farmers about sustainable, organic techniques and uncommon or forgotten food crops. I am interested in exploring strategies that will help everyone to grow and consume healthy food in the twenty-first century.

Maybe one of the most significant differences between our mission and my grandparent's profession is that they did not need to be certified or defined as "sustainable," or "organic." There was no need for organic certification in their world as some farmers use it today. In their generation, farms were largely self-sufficient and sold crop surpluses within their local communities or regional markets. There was no need for food safety inspections because if you made your neighbors sick with bad vegetables, word would spread quickly and punishment would be swift and long-lasting.

The rise of global trade networks, high-yielding industrial agriculture, and improved transportation is an age where petrochemicals are inexpensive and accessible to all, even home-owners. This made the first proponents of the organic movement speak up. They said "no" to the global marketplace and megacorporations and reexamined some of the ideals formed through centuries of experiences of small-scale, sustainable farming communities. These practices are based on careful scientific observations and an understanding of ecology. The land, soil, and people are taken care of rather than exploited. The first organic farmers made a political statement that they wanted farming to be socially as well as ecologically productive and lasting for generations to come.

My grandparents did not see a need for organic because they did not spray chemicals except on rare occasions. They remember that agriculture was less of an industry and more of a family and community endeavor. Community was important because fresh could not be imported in December. The community had to take care of their soil and local food supply.

The global marketplace began to catch up with my grandparents, and they retired right before the first inexpensive flower imports from South America, with dirt-cheap blooms laden with chemical sprays, began flooding local markets. By that time my parents already worked jobs off-the-farm.

I had to relearn many of the lessons that my grandparents learned through a lifetime of experience as I tended this abused soil that had virtually no biological life. It would not even hold enough water to germinate sensitive seeds like carrots. Just like the first proponents of organic agriculture, I made a choice to say "no" to senseless consumption and to begin to build on the legacy of my forefathers, adapting their wisdom to modern times. And my grandpa was there in the early years, christening our new property with a ten-pound sack of turnip seeds. He spread them for days to improve the health of the land, joking that he thought the bag "would never get all." Symbolically, he was preparing the ground for future generations.

The History of the "New" Organics

Joseph Heckman and Mark Keating

Organic agriculture in its present form evolved from over a century of scientific and spiritual inquiry based on the philosophy that nature provides the optimal model for an efficient and enduring crop and livestock production (Confort 2001; Lotter 2003; Tredwell et al. 2003; Heckman 2006; Lockertetz 2007). As such, organic farming began as a direct refutation of the mechanistic understanding of agriculture that emerged from the Industrial Revolution. That model discounted the significance of ecological principles in agriculture by likening living organisms to complex machines and managing soil as a chemical matrix. The emergence of the mechanistic model led to the use of manufactured chemical fertilizers, causing a displacement of animal manures and cover crops and the substitution of pesticides for cultural pest management practices. It was also characterized by declining farm biodiversity as integrated crop and livestock production operations gave way to monocultures. Many agriculturalists saw great promise in the increased productivity the industrial model appeared to offer, even more so after the advent of fossil fuels and the mechanization of farm labor led to greater scale of production. However, the early proponents of organic agriculture detected fundamental deficiencies within industrial farming systems, which they determined could only be corrected through a more holistic, ecologically based approach.

While it represents a direct challenge to the industrial paradigm, it would be a mistake to conclude that organic agriculture simply reflects farming as was practiced before synthetic fertilizers, pesticides, and motorized farm machinery became available. In fact, the soil scientists who first identified organic agriculture voiced alarm that the traditional agricultural practices then common to Europe and the United States were inherently unsustainable. By the early 1900s, these scientists had identified a connection between decreases in agricultural productivity and declining soil fertility, including rapid soil loss due to erosion. They recognized a historical pattern in the United States, for example, in which farmers migrated from historical farming regions that suffered "soil exhaustion" to newly available land to the south and west. Like their academic peers who were at the time advancing the industrial model, the organic proponents sought to overcome the limitations of traditional agriculture as was practiced in the West. What distinguished them was their skepticism that the industrial model could provide a lasting alternative and their willingness to look further for an approach that would.

Franklin King, a renowned soil physicist from the University of Wisconsin, who also attained a senior position at the USDA, foresaw the impermanence of Western agriculture when he visited Asia in 1909. Traveling extensively for nine months, he observed that the peasant farmers were working on land that had remained productive far longer than the American farmland would (Heckman 2013). The title of his 1911 book that describes his discoveries—*Farmers of Forty Centuries or Permanent Agriculture in China, Korea, and Japan*—conveys sustainability without using that modern term. King described how Asian farmers utilized every natural by-product arising from their daily lives, including livestock and human manures, crop residues, ashes, and sediment from canal dredging, as soil amendments. There was literally no waste within their farming systems because all organic matter was scrupulously returned to the land. *Farmers of Forty Centuries* was thoroughly documented with pictures of compost piles, intercropping, and other traditional Asian farming practices and set the course for future investigation. The book enshrined the practice of rigorously returning fertility to the soil through the application of natural materials, including manure, and the maintenance of continuous ground cover as the fundamental principles of organic farming.

Sir Albert Howard along with his wife and research collaborator Gabrielle Matthei similarly discovered remedies to the deficiencies of Western agriculture from their observations of the practices of Asian peasant farmers (Gieryn 1999). Born in 1873 and raised on a traditional crop and livestock farm in England, Sir Albert was instinctively skeptical about the modernization of agriculture. He recognized that the replacement of draft animals with machinery and that of manure with synthetic fertilizers would inevitably degrade soil quality and diminish its microbiological vitality. After they both graduated from Cambridge University with honors, they spent more than twenty-five years as accomplished soil scientists and plant breeders at research facilities across India and modern-day Pakistan. Throughout their travels, they observed how diligently native farmers compiled, managed, and applied compost on their farmland and derived great benefits from this practice.

The Howards distilled the art and science of composting into a standardized protocol, which they called the Indore Composting System, named after the region where they developed the process (Howard 1943; Fitzpatrick 2005). In the Indore Composting System, plant- and animal-derived feedstocks of high carbon-to-nitrogen (C/N) ratios were layered with materials of lower C/N ratios into large piles. By managing oxygen and moisture levels and core temperature within the pile, nutrient-rich compost virtually free of pathogens and weed seeds was produced. After Gabrielle's passing, Sir Albert compiled the fruits of their collaboration in a work entitled *An Agricultural Testament*, which he dedicated to her. The book provides detailed guidance on the Indore Composting System and stresses on the Rule of Return—Howard's dictum that all nutrients removed through crop harvest must be restored through the application of composted plant and animal materials and cover crops.

The Howards were also early proponents of the theory that foods raised on organically managed soils would be nutritionally superior to those raised otherwise (Howard 1972). They attributed the enhanced benefit to the abundance of mycorrhizal fungi found in soils whose biological activity is sustained by compost, crop rotations, and cover crops. They surmised that soluble, protein-rich compounds found in mycorrhizal fungi imbued organically raised plants with exceptional physical characteristics, including resistance against disease. The Howards believed that the nutritional attributes of organically managed crops could be transmitted through the food web to livestock and humans. They argued that disease—whether in plants, livestock, or humans—is a manifestation of nutritional deficiency and that restoring soil fertility would provide "the public health system of tomorrow." In 1945 Sir Albert characterized their findings thus, "If I were asked to sum up the results of the work of the pioneers of the last twelve years or so on the relation of agriculture to public health, I should reply that a fertile soil means healthy crops, healthy livestock, and last, but not least, healthy human beings" (https://web.extension.illinois.edu/smallfarm/downloads/51475.pdf).

The Howards' hypotheses on the nutritional attributes of food raised on humus-rich fertile soil coincided with more scientifically validated findings from two of their peers (Howard 1946; Balfour 1976). Sir Robert McCarrison, a Scottish physician, conducted extensive research in the Hunza region of India, which he summarized in his book *Nutrition and Health*. The Hunzas followed the Rule of Return and produced virtually all of their food on mineral-rich soils fertilized with compost. McCarrison documented the exceptional fitness maintained by the Hunzas and contrasted it with the worsening health of Europeans where food from industrial agriculture was increasingly the norm. American dentist Weston Price traveled the world during the 1930s and 1940s to study more than a dozen isolated populations of indigenous people who maintained their traditional agricultural and dietary practices. Without exception, he observed vibrant health without physical or mental defect. Price repeatedly observed that whenever these indigenous peoples changed over to commercialized modern foods such as refined flour products, sugar, canned good, and polished rice, they developed dental caries, deformed skeletal structures, and declining health overall. Dr. William Albrecht, a soil scientist and seminal influence on organic soil management, contributed a chapter entitled "Food Is Fabricated Soil Fertility" to Price's milestone book *Nutrition and Physical Degeneration*.

These accumulating observations and ideas on food and farming caught the attention of Lady Eve Balfour in England and Jerome Rodale in the United States. Balfour was known as the "Voice of the Organic Movement" with her speeches, travels, and writings (Brander 2003). Her bestselling book, *The Living Soil*, was a digest of the emerging alternative agriculture movement that would become organic farming. In later editions of her book she reported on a long-term side-by-side comparison between organic farming and conventional chemical-based farming system. In 1946, Balfour founded The Soil Association to explore "a direct connection between farming practice and plant, animal, human and environmental health." In 1967 this organization became the first in the world to establish organic production standards with a corresponding certification program.

Jerome Rodale began his family's continuing association with organic agriculture in 1942 by introducing *Organic Farming and Gardening* magazine, which featured Sir Albert Howard as a frequent contributor (Hershey 1992; Tredwell et al. 2003). J. I. Rodale and his son Robert Rodale would become tenacious spokespersons for the organic movement in the United States. The Farming Systems Trial, the longest, continuous, replicated side-by-side comparison between organic and chemical agriculture farming systems, was launched in 1981 at the Rodale Institute's Research Farm in Pennsylvania. Decades into research on farming systems, the Institute concluded, "After an initial decline in yields during the first few years of transition, the organic system soon rebounded to match or surpass the conventional system." Rodale press publications on organic farming and gardening achieved wide circulation and became indispensable resources for the back-to-the-land movement.

While his contributions went further than nomenclature, it is now generally agreed that Lord Northbourne was the first person to use the word "organic" in reference to a farming system (Scofield 1986). Prior to his influential book, *Look to the Land*, published in 1940, terms such as "nature farming" and "humus farming" had prevailed. In that book, Lord Northbourne conceptualized "the farm as a living whole," or an "organic whole." He used "organic" to mean complex interrelationships of parts, similar to that of living things. Lord Northbourne wrote that "the farm itself must have a biological completeness; it must be a living entity, it must be a unit which has within itself a balanced organic life." He added that should a farm depend upon imports for soil fertility, it "cannot be self-sufficient nor an organic whole." In conceptualizing the "farm as organism," Lord Northbourne drew upon Ehrenfried Pfeiffer's writings on biodynamic farming, which were themselves derived from Rudolf Steiner's 1924 declaration: "Truly, the farm is an organism."

This early community of organic researchers and practitioners conceived of organic farming as the integration of cultural practices that would be consistent with how nature functioned. Key cultural practices and concepts in organic farming included the Rule of Return, building soil humus and organic matter content, composting, extended crop rotations, integration of crops and livestock production, connections between food quality and soil fertility, protecting food quality from industrial processing, and rejection of inappropriate or synthetic technologies that displace natural processes. Sir Albert Howard perhaps best conveyed the basic agronomic principles in *An Agricultural Testament* this way:

> Mother earth never attempts to farm without livestock; she always raises mixed crops; great pains are taken to preserve the soil and prevent erosion; the mixed vegetable and animal wastes are converted into humus; there is no waste; the processes of growth and the processes of decay balance one another; ample provision is made to maintain large reserves of fertility; the greatest care is taken to store the rainfall; both plants and animals are left to protect themselves from disease.

By 1960, the pioneers had articulated a scientifically grounded, spiritually uplifting vision for organic agriculture, which was nonetheless largely irrelevant to the changes then transforming the countryside. The components of this transformation were not new—chemically derived fertilizers and pesticides were introduced in the nineteenth century and hybrid seeds and mechanized tractors became commercially available during the 1920s. However, conventional agriculture became part of an accelerated cultural transformation after World War II as the marriage between the Industrial Revolution and the scientific discoveries of the twentieth century became synonymous with progress. Faith in human mastery of the environment—reflected in the ability to decode DNA or travel to outer space—convinced many people that technology had no limits. New technologies and materials that appeared promising under controlled conditions were rapidly commercialized and disseminated. For example, after the insecticidal properties of DDT were identified in 1939, the pesticide was used to suppress life-threatening outbreaks of typhus and malaria. By the early 1960s, however, DDT had become a staple of commercial agriculture in the United States and was also sprayed in parks, beaches, and residential neighborhoods to combat nuisance insects. In a phenomenon that would become known as the agricultural chemical treadmill, American farmers adopted piece by piece the brave new world of industrial agriculture that academic, commercial, and governmental authorities enthusiastically endorsed.

If the pioneers gathered the kindling to fuel the modern organic movement, Rachel Carson brought the spark that ignited the flame (Beyl 1992). Carson was a talented writer and a biologist employed by the Bureau of Fisheries at the Department of Interior. She is best known for her book *Silent Spring*, published in 1962. This seminal work grew out of an alarm caused by the widespread application of persistent synthetic pesticides. *Silent Spring* documented the risks of these toxins in the food chain and their harmful accumulating effects on the interconnected web of life. This bestselling book elicited a withering rebuke from powerful interests, but it nevertheless served to fuel an environmental consciousness that strengthened the cause for organic farming.

Like the great naturalists John Muir and Aldo Leopold, Carson popularized the understanding that human beings are an intrinsic element within a natural world that we can neither dominate nor control. Following their lead, Carson warned that extremely adverse consequences would result from ignoring the principles of ecology. Rather than advocating for the prohibition of synthetic pesticides, she argued for an appraisal of the risks and benefits in specific applications. She also recommended accelerated study and utilization of systems-based biological pest management practices. Carson's powerful message was indirectly connected to the pioneers' organic paradigm: Nature, not mankind, is in charge, and everything is connected to everything else.

Carson never directly connected her message to organic farming, but the organic movement used her material to push forward with organic farming as the alternative to the pesticide problem. What distinguished *Silent Spring* and comparable works by Carson's contemporaries Murray Bookchin and Barry Commoner was their clarion call about the dangers of toxic synthetic substances in general. Organic farmers primarily thought and worked at the scale of small, self-sustaining farms, whereas Carson and her peers addressed an environmental crisis without boundaries. A shift in focus from farm-scale maintenance of optimal soil fertility to a more universal concern about the potential toxicity of synthetic substances greatly expanded public interest in organic agriculture. At the same time, it contributed to a narrower understanding of organic farming in which avoidance of any and all potentially toxic materials came to supersede the distinctly local, integrated, and self-sustaining vision of the organic pioneers as most elaborately articulated by Lord Northbourne.

The massive public attention that *Silent Spring* attracted, which included supportive comments from President Kennedy, spurred countless Americans to source—by growing their own, if necessary—food grown without synthetic chemicals. Rodale's *Organic Gardening and Farming* magazine came to surpass one million subscribers and spurred the back-to-the-land movement, which produced a new generation of innovative farmers. For philosophical inspiration, this movement drew on works such as Helen and Scott Nearing's *Living the Good Life* and E. F. Schumacher's *Small Is Beautiful*, which espoused a more self-sufficient and less materialistic lifestyle. The growing consumer demand also enabled traditional farmers who had avoided adopting the industrial model to access new markets for natural products. The Lundberg family of California, rice farmers, and Frank Ford, who raised and marketed West Texas whole wheat under the Arrowhead Mills label, built national brands by adapting traditional production practices and eschewing synthetic inputs. Charles Walters Jr., who lived through the Dust Bowl as a Kansas farm boy, began publishing the magazine *Acres, USA* in 1972, which promoted natural sources of soil fertility and cooperative economics. *Acres, USA* sponsored conferences (Ingram 2007) that focused attention on the works of many maverick scientists such as Drs. Wilhelm Reich and Phil Callahan. Reich's functional approach to the study of nature offered an alternative to the dominant machine model of life. Callahan investigated the role of subtle energies in soil and plant health. The magazine popularized the research of Dr. William Albrecht of the University of Missouri, who echoed the Howards in writing that "a declining soil fertility, due to a lack of organic material, major elements, and trace minerals, is responsible for poor crops and in turn for pathological conditions in animals fed deficient foods from such soils, and mankind is no exception."

For fifteen years following the publication of *Silent Spring*, anecdotal successes in the field and budding consumer demand did nothing to soften USDA's strident opposition to organic agriculture. In 1974, Secretary of Agriculture Earl Butz commented that transitioning American agriculture to organic production would result in starvation for fifty million citizens. At the time, Secretary Butz was championing an export-driven agricultural economy in which ever-larger farms ("Get big or get out," he advised) would drive up commodity production by planting fields "fence row to fence row." This accelerated industrialization of American agriculture was financed by lending practices that greatly expanded the availability of farm credit. It also increased dependence on fossil fuels used to manufacture and apply synthetic fertilizers and pesticides. Under the right growing conditions this model achieved significant yield increases, but spikes in interest rates and energy prices in the late 1970s shook the agricultural economy across America. The crisis that ensued in the countryside—several hundred thousand family farms would exit agriculture in the decade that followed—finally opened the USDA's eyes to organic agriculture's potential.

Bob Bergland, a Minnesota farmer and state legislator who served as Secretary of Agriculture in the Carter administration, had a neighbor who was a successful organic farmer (Youngberg and Demuth 2013). The mounting economic distress among family farmers combined with a 1977 USDA study that documented alarming trends with soil, water, and wildlife resources on private lands, along with conversations with organic farmers across the fence convinced him that organic production merited closer consideration. Dissatisfied with the USDA's grasp of the subject, Secretary Bergland appointed a study team of senior department officials coordinated by Dr. Garth Youngberg to go straight to the source. Working for a year, the study team compiled sixty-nine case studies of working organic farms in twenty-three states. The study team also drew upon results from a detailed survey of more than 700 magazine subscribers conducted by Robert Rodale, who had succeeded his father at the publishing house and research farm.

In July 1980, the study team's largely favorable findings were published as the *Report and Recommendations on Organic Farming*, which quickly became the USDA's most frequently requested publication. The *Report* noted that organic farmers were not regressing to preindustrial agricultural practices but rather used "modern farm machinery, recommended seed varieties, certified seed, sound methods of organic waste management and recommended soil and water conservation practices." The *Report* characterized organic farms as "productive, efficient and well-managed." Stepping lightly on sensitive ground, the study team stated in its summary of major findings that "there are detrimental aspects of conventional production, such as soil erosion and sedimentation, depleted nutrient reserves, water pollution from runoff of fertilizers and pesticides, and possible decline of soil productivity. If costs of these factors are considered, then cost comparisons between conventional (i.e., chemical intensive) crop production and organic systems may be somewhat different in areas where these problems occur." This conclusion subtly referenced the "true cost of food" concept that holds that industrial agriculture appears more profitable than organic production only because the externalized consequences of its adverse environmental impacts are not accounted for.

The study team's definition of organic farming reflects the era's shift in focus from whole farm to individual materials:

A production system which avoids or largely excludes the use of synthetically compounded fertilizers, pesticides, growth regulators, and livestock feed additives. To the maximum extent feasible, organic farming systems rely upon crop rotations, crop residues, animal manures, legumes, green manures, off-farm organic wastes, mechanical cultivation, mineral-bearing rocks, and aspects of biological pest control to maintain soil productivity and tilth, to supply plant nutrients, and to control insects, weeds, and other pests.

Reflecting Carson's perspective, the definition began by stating what organic farming is *not*— namely, a place where synthetic materials are used. The definition made the pioneers' emphasis on natural practices and materials secondary, though the case studies certainly reinforced the importance of the farm as a self-sustaining operation. In total, the *Report* presented a thoughtful scientific and economic validation of organic farming and concluded that it could represent a viable commercial alternative for family farms. Political headwinds, however, stymied any immediate impact. The incoming Reagan administration rejected further formal consideration of organic agriculture and, to illustrate the point, terminated Dr. Youngberg's employment.

In the following year, 1981, professional societies of soil scientists and agronomists organized a symposium on organic farming (Bezdicek et al. 1984). An objective was to explore how organic farming may contribute to a more sustainable agriculture. A publication that followed concluded that organic farming most definitely could make a contribution. It also noted that "the soils for the two farming systems may be quite different, each with its own unique chemical and biological properties and crop production capabilities." During the 1980s interest in organic agriculture was limited in agricultural universities, but a few scientists continued to quietly work on organic research projects. As a way to find a more appealing term, the sustainable agriculture program was initiated as an alternative to organic agriculture. This program eventually evolved into the USDA national competitive grants program now known as SARE. Under the SARE program many research projects were conducted that were supportive of organic farming (Grubinger 1992).

The American organic movement made progress during the 1980s in association with the growth of organic certification (Obach 2015; Gurshuny 2016). Through this process, a certifying agent attested that the products from the certified operation were grown and handled in accordance with a set of prohibited and restricted practices, known as standards. The Maine Organic Farmers and Gardeners Association introduced the earliest grower-based certification program in 1972, and other programs were soon operating on the West Coast and Midwest. Production standards, which were limited to crops in the early years, covered core elements, including land eligibility, soil fertility and pest management practices, protective buffers and specific allowances, and restrictions for material usage. Certification allowed organic farmers to expand their distribution network by creating a system to preserve the organic identity of products shipped to grocery store shelves and food processing facilities. These programs were not purely impartial since they often relied on farmers to evaluate each other, but the driving force was to raise, not lower, the bar. Consumers were quick to identify and support the certification programs they trusted and sought out those products bearing the corresponding seals on product labels. By the end of the 1980s, there were approximately thirty private and state government–operated organic certification programs operating across the United States. Outside the United States, the International Federation of Organic Agriculture Movements (IFOAM) was founded in 1972.

The significant commercial progress organic farming enjoyed during this period went largely unnoticed until February 1989 when the Natural Resources Defense Council published *Intolerable Risks: Pesticide Residues in Our Children's Food*. Employing novel risk assessment methodology, the report highlighted the potential harm to children aged six and under from exposure to pesticide residues commonly found on fruits and vegetables. Covering the report's release, the TV news program *60 Minutes* described one material—plant growth regulator Daminozide (trade name Alar)—as "the most potent cancer-causing agent in the food supply today." The segment reached an audience of forty million and instantly reignited the public debate surrounding pesticide safety and the benefits of organically produced foods. Nationwide, the demand for organic fruits and vegetables spiked and quickly exceeded the available supply.

The organic community—a broad term encompassing those who raised, certified, marketed, and publicly championed organic food—saw significant risk from these developments. They knew from experience that considerable time and effort would be required for the supply of organic food to meet the elevated demand. What if opportunists using watered-down standards or outright misrepresentation rushed in to supply the new markets? The separate certification programs, each with slightly different standards, were already complicating the labeling and marketing of organic foods. Each of the core constituencies in the organic community—the traditional farmers who had sidestepped the chemical treadmill, the younger generation of back-to-the-land farmers, and Rachel Carson's environmentalist protégés—were instinctively skeptical of the USDA. However, after the explosion of interest following the Alar story, they reached a general consensus that a single federal certification program would be necessary to clarify and protect the public understanding of organic farming.

Senate Agriculture, Nutrition, and Forestry Committee member Patrick Leahy of Vermont had already been pursuing a federal organic certification program for several years. In June 1990 he introduced the Organic Foods Production Act (OFPA) drawn from a lengthy collaboration between the organic community and committee staffer Kathleen Merrigan. Leary that USDA would reject outside input once the program became operational, the organic community insisted that OFPA contain checks-and-balances to preserve its own seat at the table. The most significant of these was the creation of the National Organic Standards Board (NOSB), a fifteen-member body appointed by the Secretary of Agriculture to include farmer (4), processor (2), consumer (3), environmentalist (3), certifying agent (1), scientist (1), and retailer (1) representatives. Charged with advising the secretary on organic standards and policy, the NOSB was specifically authorized to determine which synthetic substances could be used by certified crop and livestock operations. The Secretary retains final authority for authorizing such substances but may only choose from among those favorably recommended by the NOSB.

A second check-and-balance provision in the OFPA had a lasting effect on preserving the organic community's influence on the USDA certification program. While the law authorizes USDA to establish the organic crop, livestock, and handling (processing) standards, it delegates the day-to-day compliance duties to accredited, third-party certifying agents. From developing an organic system plan through the on-site inspection to the annual renewal process, certified farmers and handlers work directly with the certifying agent of their choice. The majority of the preexisting private and state government certification programs would eventually become USDA-accredited certifying agents and carry on serving their existing clients, along with newcomers. This provision facilitated a smooth transition to the new federal organic standard by preserving the subject expertise and client relationships that the experienced certifying agents brought with them. Utilizing third-party certifying agents continues to foster a competitive market place and provides certified farmers and handlers a choice of whom to work with. It also spares USDA from having to hire and train a sizable in-field compliance staff of its own.

While the organic community was largely on board, support for OFPA within Washington, DC, was far from universal. The House Committee on Agriculture refused to even consider legislation authorizing an organic program, and USDA's leadership showed no interest of its own. The impasse was broken during the House of Representatives floor debate on the 1990 Farm Bill, formally named the Food, Agriculture, Conservation and Trade (FACT) Act. Congressman Peter De Fazio of Oregon proposed amending the House FACT draft by incorporating the OFPA, which the Senate had already passed, in its entirety. Bypassing the Agriculture Committee leadership was a bold move for a junior congressman who did not even sit on that panel, but a groundswell of grassroots support carried the day. The House of Representatives approved OFPA, which became the only floor amendment added to the Farm Bill that year. The OFPA was signed into law by President George H. W. Bush on November 28, 1990. The victory heralded Kathleen Merrigan's emergence as a key Washington, DC, ally for the organic community—a role she would reprise frequently in the following years.

The momentum generated by Congress slowed once the workload shifted to USDA. The NOSB began meeting and issuing recommendations in 1992, but the Department made little internal progress before hiring Grace Gershuny in 1994. A veteran of the Northeast Organic Farming Association, she had been instrumental in drafting standards and certification procedures for private certification programs in the region. Combining NOSB recommendations with her own experience, Gershuny pulled together USDA's first proposal for a comprehensive certification program that was released for public review in December 1997. To USDA's shock, the proposal generated more than 280,000 public comments, which almost unanimously vilified the effort as a fatally flawed abandonment of organic tradition. While little in the proposal escaped criticism, the predominant concerns expressed were that USDA lacked a holistic understanding of organic farming and had failed to prevent the intentional and inadvertent presence of prohibited substances in organic farming.

While the USDA proposal suffered from discrete flaws, the backlash also reflected public misunderstanding about organic certification itself. Lacking direct experience with the process, many in the public had let idealistic assumptions shape their perceptions about the certification's breadth and rigor. For example, commenters criticized the provisions for managing and applying compost as excessively lax, yet the simple procedures USDA proposed were consistent with historical organic standards. Similarly, requirements for a variety of practices, including crop rotation and livestock confinement, were less a dilution of earlier standards than a reflection of the flexibility farmers needed to adapt to changing conditions. By and large, commenters did not appreciate that certifying agents and farmers had traditionally worked out compliance based on the farm's specific conditions with an eye toward making improvement over time.

An even greater source of friction was the public's perception that USDA was inadvertently or worse yet intentionally sanctioning a flood of synthetic substances into organic farming. These concerns were fueled by an inaccurate belief that the USDA proposal sanctioned the use of genetic engineering, ionizing radiation (an FDA-approved food sanitation practice), and the application of sewage sludge on organic land. Consumers had long seen organic certification as a firewall against any synthetic substances in their food, and headlines about potential contaminants—especially the "Big 3"—caused an uproar. In fact, the USDA proposal did not sanction any of the "Big 3" and expressly prohibited a wide variety of synthetic substances, including toxic inert ingredients in pesticides. Much to the public's confusion and consternation, however, the proposal stated that there were other incidental synthetic substances on organic farms that simply needed to be accepted without review.

How best to approach these incidental synthetic substances revealed another gap in the public understanding of organic certification. While the pre-federal organic certification programs prohibited synthetic substances other than those specifically allowed (such as vaccines), their compliance and enforcement activities had limitations. For example, they could review each agricultural ingredient in livestock feed for compliance, but what about the dozens of additives such as vitamins, minerals, preservatives, and flow agents that were routinely added? Similarly, how could a certifying agent identify and assess the ingredients in a carrying material such as the ointment used to deliver the therapeutic agent in a livestock medication? For the sake of practicality, USDA proposed a general allowance for certain functional synthetic materials, but the vast majority of commenters insisted upon rigorous procedures for identifying and eliminating even trace levels of any which were not expressly approved.

Responding to the overwhelming public pressure, Secretary of Agriculture Dan Glickman pledged to issue a revised proposal that the organic community could embrace before the conclusion of the Clinton administration. USDA poured resources into the effort, including reconvening the NOSB on a frequent basis and adhering more closely to its recommendations. After a second proposal garnered an additional 45,000 public comments, USDA issued the regulations for an organic certification program (formally named the National Organic Program [NOP]) on December 21, 2000. The organic community largely agreed that Secretary Glickman had kept his word by delivering more prescriptive crops and livestock standards and retaining the fundamental working relationship between certifying agent and farmer. The regulations firmly embraced the more materials-centric focus, including a categorical prohibition on materials derived from genetic engineering (defined as an "excluded method"), which the public had come to equate with organic certification.

The establishment of the NOP was a historic achievement in organic farming's quest for acceptance by the agricultural establishment, but USDA remained measured in its embrace. Most significantly, the Department denied that organic production and handling practices imparted any nutritional, food safety, or even flavor attributes to certified food. Speaking at the NOP's launch, Secretary Glickman stated, "Let me be clear about one other thing. The organic label is a marketing tool. It is not a statement about food safety. Nor is 'organic' a value judgment about nutrition or quality." He acknowledged that under certain circumstances organic farming could provide greater conservation benefits than conventional practices. However, he unequivocally reaffirmed USDA's commitment to industrial agriculture in general and genetic engineering in particular by citing the importance of those practices to "feeding the world." It remains the USDA's official position that organic certification simply establishes that an accredited certifying agent has affirmed that the certified product was produced and handled in accordance with the standards and that no qualitative difference should be implied.

Regardless of USDA's position, market research before and after the NOP has consistently established that consumers who purchase organic products are primarily motivated by perceived nutritional benefits, most significantly, lower levels of pesticide residues. Over the years, results from the USDA's Pesticide Data Program, which annually tests an array of conventional and organic food items, have consistently demonstrated that the certified items actually do contain fewer residues. While residues from the limited number of pesticides allowed under the organic standards are found on some certified items, they tend to be fewer in total than the number of detections on conventional products. There is no question that far more pesticides are allowed in conventional production and that more are used, frequently with multiple products applied to a single crop. The Environmental Protection Agency, which is responsible for establishing allowed pesticide residue levels on food, along with USDA are each quick to point out that the vast majority of Pesticide Data Program samples show either no detection or detection at a level well below that determined to be safe.

While the data on pesticide residues is straightforward, organic certification in and of itself provides a less reliable indicator of a food's nutritional quality. Modern research has produced an abundance of evidence validating the principle that how plants and animals are raised will reflect their nutritional content. Much of this research also supports the organic pioneers' conviction that practices such as optimal soil fertility, raising livestock on pasture, and providing them a species-appropriate natural diet contribute directly to beneficial nutritional attributes. However, it is not possible in every instance to extrapolate from the general principles of organic farming to draw conclusions about specific certified products. There can be many variables in the production process over which organic certification has little or no control, as well as variation between the practices on different certified operations. The great value of organic certification is that it verifies basic but meaningful differentiation between organic products and the conventional food supply. However, it would be unrealistic to use organic certification under all circumstances as a pure proxy for organic farming.

One promising outcome of establishing a consistent national standard has been the increased interest in organic farming from agricultural research institutions. A groundbreaking report from the Organic Farming Research Foundation in 1997 established that less than one-tenth of 1 percent of USDA's research portfolio had "strong" relevance to organic production. With the NOP in place, Congress used the 2002 Farm Bill to authorize two dedicated funding sources—the Organic Agriculture Research and Education Initiative, and the Organic Transitions Program. The research these programs fund must incorporate organic stakeholders in experimental design with all fieldwork using certified land and/or livestock. Through 2014, USDA had allocated more than $140 million through these programs. Organic agricultural researchers are now applying sophisticated analytical techniques to fundamental elements such as the synergistic benefits of dynamic soil microbiology, the interactions between soils, plants, and livestock living communally. Applying traditional plant breeding practices—not genetic engineering—to develop new plant varieties and livestock breeds adapted to local production conditions has also received significant funding.

Looking back in time at the decision to unify organic certification under USDA, proponents can point to multiple benefits of having done so. In addition to the growth in funding for organic agricultural research, USDA now provides a wide variety of services to the organic community. For example, USDA provides an annual certification cost share to producers and handlers, which reimburses 70 percent of their certification costs, up to a limit of $750. The USDA's Natural Resource Conservation Service offers certified and transitional farmers' payments for environmentally beneficial practices such as cover cropping and pollinator habitat protection, though funding is capped far below the amount conventional farmers can receive. In 2010, the livestock standards were significantly enhanced to require that ruminants derive at least 30 percent of their forage needs from pasture during the grazing season. Consistent growth in the organic sector appears to validate USDA's stewardship. The number of certified farms, which USDA estimated to be 6,592 in 2000, grew to 12,818, encompassing 4.4 million acres by 2015. The USDA estimated the farm gate value of organic sales from these farms at $6.2 billion.

Despite the steady growth in organic farms, acreage, and sales, some in the organic community object to what they see as the industrialization of organic production that USDA has sanctioned. These critics maintain that due to ambiguous standards and poor enforcement—especially with the livestock standards—the largest certified farms are little different from conventional operations. In recent years, the certification of massive greenhouse operations raising produce in potting mix with liquid fertility sources has garnered complaints that being organic no longer even requires soil. Doubtless, the organic pioneers would do a double take were they to see how the historic practices and principles they championed have been adapted to contemporary production systems. Perhaps, though, they would recognize that today's world is sufficiently different from their own to justify accommodation for scale and efficiency. Either way, they could feel satisfied that their discoveries and insight have inspired so many others to farm in closer harmony with nature.

References

Alternative Farming Systems Information Center. Oral History Interview Series. www.nal.usda.gov/afsic/ (accessed May 19, 2005).

Balfour, E. B. 1976. *The Living Soil and the Haughley Experiment*. New York: Universe Books..

Beeman, R. 1993. The Trash Farmer: Edward Faulkner and the Origins of Sustainable Agriculture in the United States, 1943–1953. *Journal of Sustainable Agriculture* 4: 91–102.

Beyl, C. A. 1992. Rachel Carson, *Silent Spring*, and the Environmental Movement. *HortTechnology* 2: 272–5.

Bezdicek, D. F., Power, J. F., Keeney, D. R., and Wright, M. J. (eds). 1984. *Organic Farming: Current Technology and Its Role in a Sustainable Agriculture*. Madison, WI: American Society of Agronomy, Crop Science Society of America, Soil Science Society of America..

Brander, M. 2003. *Eve Balfour, The Founder of the Soil Association and Voice of the Organic Movement*. Haddington: Gleneil Press.

Carson, R. 1962. *Silent Spring*. Boston: Houghton Mifflin.

Conford, P. 2001. *The Origins of the Organic Movement*. Glasgow: Floris Books.

Congressional Record. 1990. Public Law 101–624, 28 November 1990. Food, Agriculture, Conservation and Trade Act of 1990. Title XXI, Organic Certification. Congressional Record S10959. Washington, DC.

Fitzpatrick, G. F., Worden, E. C., and Vendrame, W. A. 2005. Historical Development of Composting Technology during the 20th Century. *HortTechnology* 15: 48–51.

Fromartz, S. 2007. *Organic, Inc.: Natural Foods and How They Grew*. Orlando: Harvest Books.

Gieryn, T. F. 1999. *Cultural Boundaries of Science: Credibility on the Line*. Chicago: University of Chicago Press. pp. 233–335.

Grubinger, V. 1992. Organic Vegetable Production and How It Relates to LISA. *HortScience* 27: 759–60.

Gurshuney, G. 2016. *Organic Revolutionary: A Memoir of the Movement for Real Food, Planetary Healing, and Human Liberation*. Vermont: Joe's Brook Press.

Heckman, J. R. 2006. A History of Organic Farming: Transitions from Sir Albert Howard's War in the Soil to USDA National Organic Program. *Renewable Agriculture and Food Systems* 21: 143–50.

Heckman, J. R. 2013. Soil Fertility Management a Century Ago in Farmers of Forty Centuries. *Sustainability* 5: 2796–801. www.mdpi.com/2071-1050/5/6/2796

Hershey, D. R. 1992. Sir Albert Howard and the Indore Process. *HortTechnology* 2: 267–9.

Howard, A. 1943. *An Agricultural Testament*. New York: Oxford University Press.

Howard, A. 1946. *The War in the Soil*. Emmaus, PA: Rodale Press.

Howard, A. 1972. *The Soil and Health*. New York: Schocken Books.

Ingram, M. 2007. Biology and Beyond: The Science of "Back to Nature" Farming in the United States. *Annals of the Association of American Geographers* 97: 298–312.

Kelly, W. C. 1992. Rodale Press and Organic Gardening. *HortTechnology* 2: 270–71.

King, F. H. 1911. *Farmers of Forty Centuries*. Emmaus, PA: Rodale Press.

Lockeretz, W. 2007. *Organic Farming: An International History*. Oxfordshire: CABI.

Lotter, D. W. 2003. Organic Agriculture. *Journal of Sustainable Agriculture* 21: 59–128.

Mergentime, K. 1994. History of Organic. www.ofrf.org/press/otherreports.html (accessed May 19, 2005).

Obach, B. K. 2015. *Organic Struggle: The Movement for Sustainable Agriculture in the United States*. Cambridge, MA: MIT Press.

Rodale, J. I. 1942. An Introduction to Organic Farming. *Organic Farming and Gardening* (May).

Rodale, J. I. 1946. *Pay Dirt*. Emmaus, PA: Rodale Press.

Rodale, J. I. 1984. *The Organic Front*. Emmaus, PA: Rodale Press.

Scofield, A. M. 1986. Organic Farming—The Origin of the Name. *Biological Agriculture and Horticulture* 4: 1–5.

Treadwell, D. D., McKinney, D. E., and Creamer, N. G. 2003. From Philosophy to Science: A Brief History of Organic Horticulture in the United States. *HortScience* 38: 1009–1013.

US Department Agriculture. 1980. *Report and Recommendations on Organic Farming*. USDA. p. 94.

Walz, E. 2004. *Fourth National Organic Farmers' Survey: Sustaining Organic Farms in a Changing Organic Marketplace*. Santa Cruz, CA: Organic Farming Research Foundation.

Youngberg, G. and Demuth, S. P. 2013. Organic Agriculture in the United States: A 30-year retrospective. *Renewable Agriculture and Food Systems* 28: 1–35.

2

Profile: Leigh Bush, Food Anthropologist

Hello there, readers! My name is Leigh, and I am an anthropologist specializing in Food Studies at Indiana University. I've always loved food, and that passion has helped me get to know people throughout my life and throughout the world. I've eaten fried locusts in Burma and lemon flavored ants in the Amazon. I've shoveled mountains of poop on a dairy farm in Spain and loaded pitchforks full of hay in Macedonia. I'm grateful to be able to share some of the perspectives I've gained through my experiences, travels, and education as I continue my daily and lifelong exploration of how food connects us to our bodies, our community, and our environment.

It all starts when I decide to escape the inferno that is Shanghai in mid-August for the balmier surroundings of southwestern Yunnan Province. When I arrive in the small historic town called Shaxi I have no idea what day of the week it is, and haven't for a month. I've scrounged up a few new friends from my hostel stays along the way, and as we pace down the cobblestone streets toward the converted horse stable–turned quaint hostel, I spot the man. A mere two hours earlier he had been a town barber, but now his storefront is transformed. Men and women crouch around an old beam balance scale, inspecting something under the dim light. As I lean in, a few local Yunnanese look up at me with suspicion. And then I see them: mushrooms! Our friendly hostel owner, Shirley, turns to me saying, "Bai and Yi women can earn almost their entire year's income during the mushroom season here." "Do you think we might find some?" I ask. "We can try," she says.

We wake up at 9:00 a.m., beginning our journey along rural cobblestoned roads that are flanked by irrigation ditches dividing the surrounding rice paddies. As we venture upwards, the earth turns from sandy grit to red clay and underbrush begins to coat the ground, wilderness taking over. My head is glued to the ground in search of fungus. Soon enough I see one little phallus poking up from the undergrowth and just as I pluck it I see another, and another, each mushroom different from the last. I start filling my arms with treasure. Then I hear what at first sounds like birds calling to each other across the steep slope of the Shibao, but as the howl carries on I come to realize that it is a group of Bai women spread across the mountain, calling out to each other as they forage.

There are many great things about being a foodie in far-off places. One of the finest is how food often becomes or supersedes language. So, even though I speak neither Chinese nor Bai nor Yii, the mushrooms speak for me. But when I spread out my hands proudly showing off my trove, the blue-clad women pick through my collection, tossing one beautiful fungus after another to the

FIGURE 2.1 *Leigh Bush, food anthropologist. Photo courtesy of Leigh Bush.*

FIGURE 2.2 *Leigh Bush with mushrooms. Photo courtesy of Leigh Bush.*

ground. I am left with three measly nontoxic fungi. Despite my failure, Yunnan province is actually home to at least 800 edible species of mushrooms, and it is because of the specialized knowledge and human labor that people have been able to harvest these species for so long. As you will discover in this chapter, at one time spiritual beliefs actually reinforced the practical benefits of ancient agricultural systems, systems that science now shows to be organic, sustainable, and robust enough to maintain dense populations over thousands of years. Unfortunately, eroding belief systems have likewise resulted in more damaging environmental practices, an effect I felt first-hand in the smoggy density of Shanghai.

When we get to the top Shirley bargains for a lunch's worth of mushrooms, which she brings to the pavilion's owner, negotiating the price for meal preparation as well. I explore the pavilion, finding the man who is feeding the wood fire beneath the woks and the women who are deftly cleaning our fungus. The women show me how to gently rub and rinse the mushrooms then how to peel off the spongey layers underneath the cap. My eyes begin to burn from the chili pepper smoke that fills the kitchen—both the famed Sichuan pepper (a province directly north of Yunnan) and a local red chili pepper. Soon, women and men bring out the dishes, filling up the table—along with the staple rice we have Yunnan's famous dry-cured pork, our foraged mushrooms, and a green vegetable I have yet to identify.

We dine overlooking the ruins of people who learned to integrate themselves with the landscape, sustaining dense populations for thousands of years by composting to eliminate pathogens and manage nitrogen availability, and cropping to preserve nutrients from the mountains down into the valleys. Their methods of using and giving back to the land have made it so that today we might

FIGURE 2.3 *Leigh Bush with a bowl of mushrooms. Photo courtesy of Leigh Bush.*

drink tea and eat mushrooms, pork and mysterious green vegetables, and still look over a lush and fertile landscape.

We traipse back down the mountain and I'm sad to be leaving Shaxi but feel blessed to have found it in the first place. As you read this chapter, consider how ancient farming practices designed with nature and according to each local environment worked to maintain balance and fertility. Like with the Bai and Yi people of Yunnan Province, traditional methods rely on specialized knowledge and human labor to develop intricate, productive, and sustainable systems based on organic farming.

Traditional Intensive Organic Farming

E. N. Anderson

Organic Farming for 10,000 Years, Industrial for 100

Until the late nineteenth century, all agriculture was organic. The term "organic farming" has come to mean farming done without the use of artificially made or factory-made fertilizers and pesticide chemicals. These were invented in the nineteenth and twentieth centuries, and were largely confined to relatively well-to-do, self-consciously "modern" farms until well into the twentieth century. A major boom in chemical and industrial farming methods took place after World War II, and continues today. In addition, animal power has been largely replaced by machinery fueled by petroleum derivatives or natural gas.

Previously, farmers had to make do with what nature provided and permitted. Fertilizer consisted of dung, green manure, food processing wastes, and other immediately available organic products. Pest control came largely from natural agents: birds, predatory insects and mites, diseases of the pest organisms, and other natural causes. Cultural techniques—crop rotation, periodic fallowing, periodic flooding, plowing crop residues into the soil, burning fields, and many others—were often even more effective at controlling pests. Many farming cultures discovered plants toxic to pests.

As Rachel Carson predicted long ago in *Silent Spring* (1962), pesticides and habitat destruction have wiped out the natural control agents, so it is now difficult to go back to organic farming or to integrated pest management (IPM; it integrates some artificial chemical use with systematic development of "organic" regulation). Fertilizers and pesticides, as well as farm machinery, are now heavily dependent on oil, as is the whole mechanized and industrialized farming order. Currently, oil production and processing is very heavily subsidized by the United States and other countries (Juhasz 2008; Anderson 2010).

Modern organic farms reach very high levels of production. A recent authoritative report (Seufert et al. 2012) finds that organic agriculture is still out-yielded by industrial agriculture, but the gap has narrowed to 3 percent in some major categories of production, and continues to shrink.

Wet Rice Agriculture

China's rice landscapes are human creations (Anderson and Anderson 1973; Anderson 1988; Elvin and Ts'ui-Jung 1998). Burning for agricultural and other reasons in the mountains led to silt washing downstream, and buildup of alluvial deposits. This restored fertility of existing fields and allowed reclamation of new ones. This was still under way in the 1960s. Leveling fields and constructing terraces, dykes, channels, and levees is a major undertaking, but once done it is established for the long term, though requiring some maintenance. Such major land modification creates what geographers call "landesque capital"—fixed capital of high value, created by changing the landscape.

Southeast China is barely in the tropics, and has a hot, wet summer and cool dry winter. Typhoons occur, bringing rain that causes flooding and erosion. Soil is naturally poor, being based on granite; in the tropics, granite weathers to a thin, acid, low-nitrogen soil. (It does, however, have plenty of potassium and appreciable phosphorus, so these were not usually limiting nutrients.)

Agriculture is based on the rivers and streams of this hilly region. Each valley was a closed system in itself, though all were integrated to some degree into larger systems. These were bounded by drainage divides, within which everything was integrated by nutrient and energy flows.

The upper parts of these drainages were regularly burned to discourage dangerous wildlife, open the land, or simply through carelessness. It was destructive of forest resources. It kept the highlands in a young stage of ecological succession, dominated by grasses, herbs, and young trees and bushes. The herbs were very often nitrogen-fixing plants (usually leguminous), and the trees brought up deeply buried mineral nutrients. A sizable amount of these nutrients would later wash downstream into the farmed area.

Next down the hills was the zone of *fengshui* groves. Fengshui is a folk science of site planning. "Fengshui" means "wind and water," and was originally a method to maximize the benefits and minimize the harms of those two elemental forces. It is based in traditional Chinese cosmology, which involves beneficial and harmful flows of *qi*, the vital essence or energy of things. This merges the folk knowledge systems of the traditional farmers with a more strictly religious belief in good and evil spirits, magical powers, and divine energies. The basic fengshui practice in the rural landscape was pragmatic and scientific in western terms. Projecting western categories of "magic, science and religion" (Malinowski 1948) on Chinese traditional farmers is a highly artificial and often misleading process, since they did not understand the world the way today's scientists do. Much that bioscientists now explain as natural process was explained by these farmers as the work of gods or spiritual energies. Some western-style analysis is necessary here to understand the more practical aspects of traditional fengshui. The groves were left because trees are the homes of good spirits, and prevent bad spirits and bad *qi* from accumulating. They also provide the more tangible benefits of firewood, construction timber, shade, leafmold for fertilizer, wind protection, and sometimes fruit. Without fengshui and spirit beliefs, they would certainly have been cut down for quick benefits; this is proved by the fate of groves in areas where fengshui belief has weakened.

Villages were sited below the fengshui groves, which often surrounded them except on the downhill side. The groves began at the back edges of the villages, and ranged from a fraction of a hectare to several hectares. The villages were situated, whenever possible, on lee slopes. They were also placed where watercourses joined and formed pools, thus ensuring regular water in drought. They were situated above, but close to, the good farmland; sprawling onto good farmland was strictly forbidden by fengshui. The villages were above the level of floodwaters.

Near the villages were terraces used to grow vegetables. These required heavy fertilization. Nightsoil (human wastes) and other village wastes, composted, could be carried easily to these terraces. The products, also heavy, could be carried back with ease. The vegetables grew above the vitally important rice land. Vegetables are typically grown in raised beds: baulks of land about six inches to a foot high, separated by ridges. Usually the baulks are wide enough for two rows of vegetable plants. The furrows provide drainage and are used for irrigating the field; sprinklers have to be used till the plants' roots are deep enough to reach water from the furrows.

Nutrients diverted to the vegetable fields eventually washed downstream, after recycling, and wound up in the rice paddies. These were constructed in level or terraced land along the streamways. Only about 10 percent of the land was level and irrigable enough to be used for them. The fields were carefully leveled. Rice plants were started in seedbeds and then transplanted into the fields; the shock of transplanting makes the rice grow better, and also the spacing could be carefully controlled. The rice grew in water; the fields were then drained for final ripening and harvest.

Food processing wastes—from production of grain, oil, sugar, alcohol, and the like—were returned to the land, along with hearth and stove ashes, old worn-out ropes and sandals and other clothing (shredded or ashed), village dirt swept off the streets, household sweepings, leaves from trees, old paper, and old baskets. Some villagers even pounded up and then composted old bricks and roof tiles that were covered with smoke or grease, and thus rich in organic material. All animal and human wastes were composted. Composting eliminates most pathogens and reduces the high availability of nitrogen, which can otherwise overfertilize and thus "burn" the crops.

An odd source of fertilizer is blue-green algae. These one-celled water organisms fix nitrogen industriously. This fact was discovered when Philippine farmers taught American experts, in the early twentieth century, that the rice grew better on the downwind side of the paddy, because the pond scum was blown there. The Americans investigated, and found that the active ingredient of the pond scum was these algae (Copeland 1924).

Downstream from the rice fields were areas prone to deep flooding but also to drying out; these were used to raise ducks and water buffaloes. Further down were areas almost permanently under deep water; these were used to grow lotuses and other water crops, and for fishponds. Finally, nutrients washed out into salt marshes, where shrimps, fish, and shellfish were caught, and then into the ocean. Here oyster beds trapped most of the nutrients; the rest nourished a large fishery. So, in the end, no particle of nutrient that started in the mountains was lost to the system. A given molecule would be recycled many times, through fields and ponds, and eventually taken up by fish that were caught and brought up to the villages. My friend Hugh Baker, reviewing my work, commented that the success of nutrient recycling in China "must make nitrogen atoms unfortunate enough to be in other parts of the world feel unloved" (Baker 1989: 661–2, in a review of *The Food of China*).

Insects that ate insects were known. Better still were control agents that could themselves be eaten. Ducks controlled rice insects. Sometimes farmers actually paid duck keepers to turn the ducks loose in the paddies; at other times the duck rearers paid the farmers. Frogs controlled other pests, snakes (a popular food) ate rats, fish ate aquatic insect pests, weeds were fed to pigs (Anderson and Anderson 1973; Needham 1986: 519–53). Sparrows controlled flies and larvae.

Rice yields in Hong Kong in those days ran 2,500 kg/ha per crop, with two crops a year being typical. Vegetables yielded several tons per hectare. Not only were all areas of farmland producing at high levels; almost all cultivable land was used. Not only villages but other construction including graves was sited above the cropland. The dykes between rice paddies could be cropped for grass and other minor items. In other parts of the Pearl River Delta and elsewhere in China, they were used to grow mulberries, whose leaves fed the silkworms that produced the major income source of the farmers.

Writing in the early twentieth century about Chinese farming in general, Frank King stated the most relevant question clearly: "We desired to learn how it is possible, after twenty and perhaps thirty or even forty centuries, for their soils to be made to produce sufficiently for the maintenance of such dense populations" (King 1911: 2). Writing before modern soil conservation came about, he was astonished that erosion had not progressed farther. After all, America's "practices by which three generations had exhausted strong virgin fields" contrasted with East Asia's fields, "still fertile after thirty [actually eighty!] centuries of cropping" (p. 48). He observed application of up to seventy tons of canal mud (greatly enriched by organic "pollution" and algae) to each acre of cropland. He noted the fact that graves were sited off the good agricultural land.

The Pearl River delta has been the subject of classic studies, especially *Integrated Agriculture: Aquaculture in South China* by Kenneth Ruddle and Gongfu Zhong (1988). Their research was done in the 1980s. In the aquaculture ponds, six different species of carp were raised. They have different food requirements, thus being complementary and not overdrawing the system. The grass carp eats grass and weeds, the silver carp algae and phytoplankton, the bighead carp small animals (zooplankton), the black carp snails and invertebrates, the mud carp invertebrates in the pond mud, and the common carp anything left over. Catfish eat pests and larger insects, and may even squirm out of the pond to eat insects on the banks. Mullet are often stocked and fed bean processing waste and the like.

As much as 270 tons of manure and fertilizer per hectare of pond may be used, as well as up to 530 kg/ha per day of grass (Ruddle and Zhong 1988: 36–40). Water plants growing on the ponds were fed to pigs. The whole system is self-contained except for nutrients from the water and from fertilizer capture. This contrasts very strikingly with aquaculture in the United States, which is basically raising carnivorous fish on fishmeal.

Under these conditions, fishponds produced up to an incredible 7.5 metric tons per hectare per year (Ruddle and Zhong 1988: 85). I never saw yields that large; ponds I studied were producing large, high-quality fish, which required thinning and less than optimal production, and thus they produced about two tons per hectare per year, plus frogs, catfish, and other incidental items.

The dikes produced a great deal of food, as well as mulberries for silkworms. Foods included soybeans, mung beans, taro, peanut, sweet potato, sorghum, maize, and so on (p. 50), as well as squash, celery, peppers, onions, chiles, radishes, black-eyed peas, ginger, bananas, and many others (pp. 69, 72). All these can be fertilized adequately with pond mud and household wastes. Some dikes were wide enough to support large fields.

Sugar was grown, and had been for perhaps 2,000 years (Ruddle and Zhong 1988: 16). Its processing waste fed to pigs and fish, and the old leaves used for shade, roofing, and fuel. Sugar requires a vast amount of processing for a rather low-value product, and in most of the world this has made it a plantation crop—only the plantation owner has the necessary capital and scale of operations. In south China, however, the Chinese commitment to small farming and fear of the power of plantation owners led to developing other systems (Mazumdar 1998). Canals allowed small farmers easy access to central processing sites. Mills were even mounted on boats and taken from one farm to another. The emphasis on small owner-operated farms greatly encouraged the labor-intensive, biological-input-rich organic agricultural systems.

Energy flows were enormous. Except for direct sunlight and river energy, they were all derived from plants grown on the land or from human labor power. Labor requirements were also enormous: 3,620.7 person-days per hectare per year (p. 92)—ten people per hectare if everyone worked every day! The fishponds alone consumed 536 of these days.

More recently, a study of raising fish in rice paddies, carried out at Zhejiang University in the famous old city of Hangzhou (Xie et al. 2011), found that raising fish in rice paddies reduces fertilizer need by 24 percent. Pesticide use is reduced 68 percent (or more) because the fish not only eat the pests but also knock them off the rice stems. Fish themselves are a much more valuable food than the rice.

The lower Yangtze River area is comparable. Wen Dazhong and David Pimentel (1986a, 1986b) studied the agriculture of Jiaxing in the mid-seventeenth century. They used the *Shenshi Nongshu* (Shen's Agriculture Book) from the period and were able to fill in missing facts from other works and modern parallels. The region is a rice-growing area in the lower Yangtze River drainage. The yields of agriculture there today are incredibly high; they noted that in 1979 the yields of cereal were 10,455 kg/ha from double-cropping rice and from planting an additional catch crop to bridge the year end. There were thus 2.5 crops per year. Catch crops included wheat, barley, rape, and beans (Wen and Pimentel 1986a: 7).

Yields in the seventeenth century are not clear for Jiaxing, but yields in other areas ranged from around 3,900 kg/ha to 84,00 per crop (Wen and Pimentel 1986a: 5). The wheat yielded about 1,300 kg/ha, rather low for China at the time, but this was a winter catch crop, not a main crop.

In the seventeenth century, no draft animals were used. There were two systems of production: compost fertilization and green manuring. Compost fertilizing required an estimated "2330 hours of labor per hectare for the growing season" (Wen and Pimentel 1986a: 3). Much of this time, 760 hours, was involved in collecting and managing the compost, which included all kinds of dung, waste, bean cake left over from oil production, and pond sediment. Green manuring required 1,650 hours, only 80 being involved in producing and collecting the milk vetch for the green manure, but it had to be grown, taking up some land and another 240 hours of labor per hectare. Ten thousand kilograms of compost or 7,500 kg of milk vetch were added to the land.

Labor inputs in all were an incredible 1.5 million kcal/ha for composting, 1.2 for green manuring; this was essentially renewable, almost all of it being human labor. A tiny bit of charcoal went into making the hand tools. Wen and Pimentel did not count the amount of labor that, at some point, went into raising the workers and teaching them how to farm. Since this form of agriculture is incredibly skill-intensive—each farmer has to have an encyclopedia in his head—the labor of teaching is not trivial, though presumably almost all learning was on the job.

The entire agroecosystem was then, and remained until very recently, a more complex one, based on silkworm raising using mulberry leaves, livestock, fishing and fish rearing in ponds, and vegetables, as well as grain and beans (Wen and Pimentel 1986b). Mulberry trees were planted on the levees between the rice paddies, or in small plantations, thus holding the soil and providing shade—those functions alone would have made the trees worthwhile, even without their vital importance for feeding silkworms.

Accounts of the lower Yangtze area in the twentieth century showed that little changed before the Communist period. The region continued to depend on rice and silk, produced under increasingly labor-intensive conditions. This economy has been the subject of some of the great classic works on China, including Lynda Bell's *One Industry, Two Chinas* (1997), Kathryn Bernhardt's *Rents, Taxes, and Peasant Resistance* (1992), Fei Hsiao T'ung's *Peasant Life in China* (1939), and Philip Huang's *The Peasant Family and Rural Development in the Yangzi Delta* (1990). From the point of view of organic agriculture, it is important to realize that the system lasted 7,000 or 8,000 years, intensifying steadily, without depleting the land. The major downside was waterborne disease; there was little that could be done in premodern times about schistosomiasis, for instance. On the other hand, unlike other premodern societies, the Chinese could deal with malaria; they discovered early that Chinese wormwood (*qinghao, Artemisia annua*) would cure it, and the active chemical, artemisinin, is now a major drug worldwide.

Recently, Chinese scientists have been advocating some form of return to a sustainable, ecologically reasonable agriculture (Shi 2010).

The south Chinese agricultural system, supplemented by a great deal of knowledge from Southeast Asian and Australian experience, lies behind the excellent and widely used system of agriculture known as permaculture, developed largely by Bill Mollison in Tasmania (see Mollison 1988; Holmgren 2012). This is a contemporary and highly productive, but organic and sustainable, system. It is spreading worldwide and may be the foundation of the needed replacement for industrial agriculture.

Southeast Asian Variants

Chinese agriculture spread to Southeast Asia in very early times. Throughout Southeast Asia, variants of wet rice agriculture occur in almost every valley and hill range. Many of these extend the general pattern in important ways. In Southeast Asia, the forested uplands are inhabited by shifting cultivators, who clear forest, burn the cut trees and brush, plant for two years (sometimes more), and move on to a new tract of forest. Such a field is called a "swidden." The fertility captured by highland forests is thus released for these fields.

Some swiddeners move rapidly and destructively over mountain ranges, but these are generally recent in-migrants confined to the poorest lands. More typical are communities that practice long-term forest fallowing: the forest grow up for ten to fifty years and then are cut again (see Conklin 1957; Spencer 1966). Some are highly intensive and sophisticated farmers, practicing tree cropping, permanent small plots, and swidden fields as part of an integrated land management strategy. (My students and I have studied such systems in Sumatra [Lando 1979], far south China [among the Akha; Wang 2007, 2008], north Thailand [Lando 1983], and elsewhere. There is also an important indigenous literature, e.g., Pinkaew 2001 on Thailand.) Swidden fields may be monocropped—usually rice in Asia, sometimes maize—but often they are a riot of crops, with up to several dozen species in a swidden (Spencer 1966).

Unless the cycle is too short, the forests regrow, often enriched by valuable fire-following trees, notably teak. Thus nutrients are released by fire without either destroying the forests permanently or making the uplands unproductive (as they generally are in China, except where shifting cultivators have long been occupants).

Many swiddeners manage the forest to restore fertility. The most direct way is by planting trees that are known to restore the fertility of the soil. Swiddens are replanted to alders in northern Burma, casuarinas in New Guinea, leguminous trees in some other places. A major forestry industry is compatible with this, especially if timber trees are planted or at least protected.

The other alternative for tree preservation, commoner in the lowlands, is tree cropping. Southeast Asia is probably the world's greatest home area for fruit trees. These often reproduce the multilayered structure of the tropical forest: coconuts and durians dominate the skyline; below them are rambutans, lychees, mangoes, and many others; below these are bananas, guavas, papayas, and other low-fruit-bearing plants; below these, at least where there is enough sun (such as at the edges of clearings), are bushes and vegetable crops. Often, shifting cultivators plant or selectively preserve fruit trees when they cut swidden fields, and thus their habitat is converted to orchards. This tradition has been encouraged in recent years by governments and nongovernmental development organizations.

Even beach sand has its specialized set of crops: tamarind, coconuts, and rose apple grow in almost sterile sand. Another beach lover, moringa (*Moringa oleifera*) is a tree whose fruit, leaves, and roots are edible, whose wood is usable for construction, and whose bark and wood are medicinal (being antiseptic)—the world's most totally usable tree.

Root crops—actually tuber and corm crops—are a specialty of the region. Taro, some yam species, and several other root crops are native. The South American crops manioc (tapioca, yuca) and sweet potatoes were introduced in recent centuries, and have become staple foods in many areas (the sweet potatoes largely in New Guinea). In mountainous regions and on the remote islands of Oceania, rice agriculture is usually replaced by these various roots.

A vast array of vegetables and root crops grows on ground too high for rice, such as low hillocks, villages, dykes, and banks. Recycling is not as extreme as it is in China, though it is similar in general. In some parts of Southeast Asia, rice varieties exist whose stems become several feet long as floodwaters deepen; eventually the rice plants may break off and float, without the seeds rotting. This allows use of lands that flood deeply. Elsewhere, instead of irrigating, rice is planted in receding floodwaters, to grow as the land dries out from annual flooding cycles (Fox and Ledgerwood 1999).

I have often been amazed in Southeast Asia by seeing clear, pure water running off steep cultivated slopes. Runoff in the United States in such situations would be deep brown with eroded soil. The better Southeast Asian farmers prevent erosion by using groves, orchards, brushy field borders, terracing, and contour cultivating, and leaving brush to grow and fill in the gullies that do form. Contrary to all received wisdom in political-economic circles, landscape changes that profoundly transform whole mountain ranges and involve huge irrigation systems covering thousands of square miles are maintained by local villages and towns and organized entirely on the basis of kinship and local leadership (Lando 1979). Far from driving centralized states, irrigation in these systems is part of a highly centrifugal social realm with a high value on local independence (Scott 2009).

In Bali, Stephen Lansing found that water flows were staggered, which made more efficient use of the water and also prevented pest populations from building up. There was never a time when all the rice was at the same stage of growth (Lansing 1987, 1991, 2006). This story has two vital lessons for organic agriculture everywhere: First, go with traditional knowledge or at least test it; second, stagger and diversify plantings to discourage pest population buildup.

Another Continent: Agriculture in Pre-Columbian Mexico

Independently of the Old World, the Americas developed intensive and sustainable agricultural systems long before Columbus.

Contemporary Yucatec Maya agriculture in Quintana Roo, Mexico is especially well studied (see Redfield and Villa Rojas 1934; Terán and Rasmussen 1993; Terán et al. 1998; Anderson 2003, 2005; Anderson and Medina Tzuc 2005). Quintana Roo is part of the Yucatan Peninsula, a limestone shelf with extremely poor soils, erratic rains, infrequent water sources, major pest problems, and frequent hurricanes.

Their system is shifting agriculture. The staple food is maize, which supplied 75 percent of calories from Classic Maya times (first millennium CE) until very recently. This was raised in swidden fields, known as "milpas" in Mexico. As in Asia, these were cultivated for two to three years, then abandoned so that fertility could be naturally restored by nitrogen-fixing plants and other biological sources. There is, however, little input from upstream; the area has few hills and even fewer streams. A few fortunate areas do have alluvial soil enriched by inflow.

Useful trees—wild fruits and nuts, medicine, palms used for thatch, firewood sources, and shade and wildlife trees—are carefully preserved. Domestic fruit trees are planted. An old milpa is a wonderful foraging ground, and not just for people: it was a supermarket for game animals. The Maya are quite aware of this, and often deliberately manage fields for wildlife production, hunting the animals for food or other uses. "Garden hunting" (Linares 1976) is quite consciously managed, with game animals actually encouraged in abandoned milpas, because it is easy to shoot them there. Sustainable hunting was quite deliberately managed and maintained until recently (Anderson and Medina Tzuc 2005), and most species of animals were often captured young and raised for meat, so that the deer, peccaries, and other game were literally farmed in spite of not being domesticated. Parrots and grackles are occasional pests, but not serious ones.

Fertile areas are multicropped, with beans climbing the maize stalks, squash sprawling over the ground, and vegetables and small fruits growing between maize rows. The beans (four species) provide nitrogen. The squash (six species) kills pests: squash plants shade out and crowd out weeds and are toxic to many insects. Other pests are usually controlled passively by the placement of the milpas. Separated by tracts of forest, they are relatively isolated from pest sources. Birds and predatory insects from the nearby forest forage in them, eating the pest insects.

Such planting requires thorough knowledge of soils. Maize is shallow-rooted and will grow luxuriously on the very shallow soils of Maya hill areas. Beans have deeper roots (about a meter) and need deeper soil. Squash roots go down as much as two meters, and thus require a very deep and well-textured soil. Maya farmers know their areas, and also test the soil depth with a digging stick.

The only real dangers are drought, flood, and "hot rains"—rains hot enough to stimulate fungal blight growth, which can wipe out whole tracts of milpas in a few days. (I wondered why people were so frightened of "hot rains" till I saw the consequences of them during my research.) These various problems prove highly destructive, but root crops such as sweet potato and manioc, plus the tree crops, survive, while wild food resources are well known. Disasters occur frequently enough to allow parents to teach the rising generation what to do. The wide range of species and varieties guarantees that something always survives.

Mexican "design with nature," forced by harsh conditions and frequent labor shortage, contrasts with Chinese highline breeding, allowed by domesticated landscapes and lavish labor supplies. Both alternatives are skill-intensive, requiring great knowledge of pests and methods of foiling them.

More intensive agriculture includes orchards of many species of fruit, and the dooryard gardens that surround almost all houses. A dooryard garden is a permanently cultivated zone, fertilized by all the household's animal and human wastes and organic refuse. It develops, over time, an extremely fertile soil. In such gardens are grown dense tree crops, dozens of species of vegetables and flavoring herbs, medicinal plants of all kinds, and domestic livestock, especially pigs, chickens and turkeys.

A typical community can draw on about 150 species of domestic and wild food plants. Some 350 plant species are used medicinally in the communities I studied. Maya herbal medicine is strikingly effective, the Maya having discovered by trial and error a vast range of anti-inflammatory, antibiotic, and antifungal plants, diuretics, anticonvulsants, astringents, soothing plants, antidiabetic teas, and other first-aid items. Few work as well as modern drug store relatives, but some actually work better, and are used to this day, even when pharmacies are available.

Beekeeping was important, and still is locally. Native stingless bees (*Melipona beechei*) were domesticated long ago. They have been largely replaced by European honeybees, now Africanized, but still occur in many places. In the old days, and often today, the native bees were grown in hollow logs. Their natural homes in large hollow trees were also provided by burning for milpas.

Overall, given the problems of infertile soil and frequent drought or damaging rain, yields are not close to Asian rice fields. Maize yielded one to two tons per hectare per crop, and since the fallow period is typically 7–15 years, that means only about 200–400 kg/ha per year for a given field. On the other hand, the whole landscape is cropped in rotation, so the situation is comparable to the wet rice system in which the rice fields were permanently cropped but took up only 10 percent of the land.

The genius of the system lies in total management. All areas are managed for production of food, timber, medicines, domestic and game animals, vines for tying, and everything else needed.

Once Again, Variants

Yucatec Maya agriculture is more sophisticated than many Native American—or Old World—traditions, but is broadly similar to the agriculture of other tropical American peoples. Locally, other intensive systems have developed. Most productive of all are the *chinampa* fields of the lakelands of the Valley of Mexico (Rojas Rabiela 1983), wetland fields rather like the wet rice fields of Asia, except that the chinampas are raised above the water level and irrigated from the surrounding canals. Farther away are the intensive fields around Lake Titicaca in Peru and Bolivia. The altitude is so high—around 3,600 m—that crops are challenged, but the lake holds heat and maintains a warm, humid zone around its shores. In valleys draining into the lake, potatoes and other crops are grown in gardens, ridged fields, terraces, or paddy-like low areas (personal observation; see articles in Lentz 2000).

A number of crops have been developed for these high-altitude situations from Chile to Ecuador and Colombia. Potatoes come originally from the highest ranges. Other high-altitude crops include other tubers such as bitter potatoes, oca, ullucu and mashwa; seed crops including quinoa and cañihua; specialized varieties of maize (barely growing along the lake); and a lupine, tarwi, bred to be very large and succulent; and various fruits and vegetables. Llamas and alpacas were also domesticated in the high Andes, from the native guanaco, a camel relative. Only Tibet has a similar high-altitude agriculture, and its crops are less diverse and nutrition-rich. Lowland South America has a number of systems similar to Southeast Asia, with swidden fields and orchards in complementarity; a particularly good description and analysis of one system in Paraguay was provided by Richard Reed (1995, 1997).

Conclusions

In short, traditional cultures with dense populations have always managed to develop intricate, productive, and sustainable systems based on organic farming. These systems still support hundreds of millions of rural people around the world.

References

Anderson, E. N. 1988. *The Food of China*. New Haven: Yale University Press.

Anderson, E. N. 2003. *Those Who Bring the Flowers*. Mexico: ECOSUR.

Anderson, E. N. 2005. *Political Ecology of a Yucatec Maya Community*. Tucson, AZ: University of Arizona Press.

Anderson, E. N. 2010. *The Pursuit of Ecotopia*. Santa Barbara, CA: Praeger (ABC-Clio).

Anderson, E. N., and Anderson, M. L. 1973. *Mountains and Water: The Cultural Ecology of South Coastal China*. Taipei: Orient Cultural Service.

Anderson, E. N., and Tzuc, F. M. 2005. *Animals and the Maya in Southeast Mexico*. Tucson, AZ: Orient Cultural Service.

Baker, H. 1989. Review of *The Food of China* by E. N. Anderson. *China Quarterly* 119: 661–2.

Bell, L. 1997. *One Industry, Two Chinas: Silk Filatures and Peasant-Family Production in Wuxi County, 1865–1937*. Stanford, CA: Stanford University Press.

Bernhardt, K. 1992. *Rents, Taxes, and Peasant Resistance: The Lower Yangzi Region, 1840–1950*. Stanford: Stanford University Press.

Conklin, H. C. 1957. *Hanunoo Agriculture*. Rome: FAO.

Copeland, E. B. 1924. *Rice*. London: MacMillan.

Elvin, M. 2004. *The Retreat of the Elephants: An Environmental History of China*. New Haven: Yale University Press.

Elvin, M., and Ts'ui-Jung, L., eds. 1998. *Sediments of Time: Environment and Society in Chinese History*. Cambridge: Cambridge University Press.

T'ung, F. H. 1939. *Peasant Life in China*. London: Routledge and Kegan Paul.

Fox, J., and Ledgerwood, J. 1999. Dry-Season Flood-Recession Rice in the Mekong Delta: Two Thousand Years of Sustainable Agriculture? *Asian Perspectives* 38: 37–50.

Holmgren, D. 2012. *The Permaculture Handbook: Garden Farming for Town and Country*. Vancouver: New Society Publishers.

Hu, S.-Y. 2005. *Food Plants of China*. Hong Kong: Hong Kong University Press.

Huang, P. 1990. *The Peasant Family and Rural Development in the Yangzi Delta, 1350–1988*. Stanford: Stanford University Press.

Juhasz, A. 2008. *The Tyranny of Oil: The World's Most Powerful Industry—and What We Must Do to Stop It*. New York: William Morrow (HarperCollins).

King, F. H. 1911. *Farmers of Forty Centuries*. New York: F. H. King.

Lando, R. 1979. *The Gift of Land: Irrigation and Social Structure in a Toba Batak Village*. Ph.D. dissertation, Department of Anthropology, University of California, Riverside.

Lando, R. 1983. The Spirits Aren't So Powerful Any More: Spirit Belief and Irrigation Organization in Northern Thailand. *Journal of the Siam Society* 71: 142ff.

Lansing, S. 1987. Balinese "Water Temples" and the Management of Irrigation. *American Anthropologist* 89: 326–41.

Lansing, S. 1991. *Priests and Programmers: Technologies of Power in the Engineered Landscape of Bali*. Princeton, NJ: Princeton University Press.

Lansing, S. 2006. *Perfect Order: Recognizing Complexity in Bali*. Princeton: Princeton University Press.

Lentz, D. L., ed. 2000. *Imperfect Balance: Landscape Transformations in the Precolumbian Americas*. New York: Columbia University Press.

Shizhen, L. 2003. *Compendium of Materia Medica (Bencao Gangmu)*. Beijing: Foreign Languages Press (Orig. 1593).

Linares, O. 1976. Garden Hunting in the American Tropics. *Human Ecology* 4: 331–50.

Malinowski, B. 1948. *Magic, Science and Religion*. Glencoe, IL: Free Press.

Marks, R. 2012. *China: Its Environment and History*. Lanham, MD: Rowman and Littlefield.

Martinez Reyes, J. 2004. *Contested Place, Nature, and Sustainability: A Critical Anthropo-Geography of Biodiversity Conservation in the "Zona Maya" of Quintana Roo, Mexico*. Ph.D. dissertation, Department of Anthropology, University of Massachusetts-Amherst.

Mazumdar, S. 1998. *Sugar and Society in China: Peasants, Technology, and the World Market*. Cambridge: Harvard University Asia Center.

Mollison, B. 1988. *Permaculture: A Designers' Manual*. Sisters Creek, Tasmania: Tagari Publications.

Needham, J., with L. Gwei-Djen and H. Hsing-Tsung. 1986. *Science and Civilisation in China. Vol. 6: Biology and Biological Technology. Part I: Botany*. Cambridge: Cambridge University Press.

Netting, R. Mc. 1981. *Balancing on an Alp: Ecological Change and Continuity in a Swiss Mountain Community*. New York: Cambridge University Press.

Pinkaew, L. 2001. *Redefining Nature: Karen Ecological Knowledge and the Challenge to the Modern Conservation Paradigm*. Chennai, India: Earthworm Books.

Redfield, R., and Rojas, A. V. 1934. *Chan Kom, A Maya Village*. Washington, DC: Carnegie Institution of Washington.

Reed, R. K. 1995. *Prophets of Agroforestry: Guaraní Communities and Commercial Gathering*. Austin: University of Texas Press.

Reed, R. K. 1997. *Forest Dwellers, Forest Protectors: Indigenous Models for International Development*. Boston, MA: Allyn and Bacon.

Rojas Rabiela, T. 1983. *La agricultura chinampera: Compilación histórica*. Chapingo, Mexico: Universidad Autónoma Chapingo.

Ruddle, K., and Zhong, G. 1988. *Integrated Agriculture—Aquaculture in South China: The Dike-Pond System of the Zhujiang Delta*. Cambridge: Cambridge University Press.

Scott, J. C. 2009. *The Art of Not Being Governed: An Anarchist History of Upland Southeast Asia*. New Haven: Yale University Press.

Seufert, V., Ramankutty, N., and Foley, J. A. 2012. Comparing the Yields of Organic and Conventional Agriculture. *Nature* 485: 229–32.

Shi, T. 2010. *Sustainable Ecological Agriculture in China: Bridging the Gap Between Theory and Practice*. Amherst, NY: Cambria Press.

Sheng-Han, S. 1973. *On "Fan Sheng-chih Shu," An Agriculturist Book of China Written in the First Century B.C.* Peking: Science Press.

Sheng-Han, S. 1974. *A Preliminary Survey of the Book Ch'i Min Yao Shu, an Agricultural Encyclopaedia of the 6th Century*, 2nd edn. Peking: Science Press.

Spencer, J. E. 1966. *Shifting Cultivation in Southeast Asia*. Berkeley: University of California Press.

Terán, S., and Rasmussen, C. H. 1993. *La milpa de los Mayas*. Mérida: Authors.

Terán, S., Rasmussen, C. H., and Cauich, O. M. 1998. *Las plantas de la milpa entre los Mayas*. Mérida: Fundación Tun Ben Kin.

Wang, J. 2007. Landscapes and Natural Resource Management of Akha People in Xishuangbanna, Southwestern China. Paper presented at annual meeting of Society of Ethnobiology, Berkeley, CA.

Wang, J. 2008. Cultural Adaptation and Sustainability: Political Adaptation of Akha People in Xishuangbanna, Southwestern China. Final report to Sumernet Foundation.

Wen, D., and Pimentel, D. 1986a. Seventeenth Century Organic Agriculture in China, Part I: Cropping Systems in Jiaxing Region. *Human Ecology* 14: 1–14.

Wen, D., and Pimentel, D. 1986b. Seventeenth Century Organic Agriculture in China, Part II: Energy Flows Through an Agrosystem in Jiaxing Region. *Human Ecology* 14: 15–28.

Xie, J., Hu, L., Tang, J., Wu, X., Li, N., Yuan, Y., Yang, H., Zhang, J., Luo, S., and Chen, X. 2011. Ecological Mechanisms Underlying the Sustainability of the Agricultural Heritage Rice-Fish Coculture System. *Proceedings of the National Academy of Sciences*. doi:10.1073/pnas.1111043108 (accessed November 30, 2011).

3

Profile: Chef Steven Eckerd

Growing up in a small town in the center of Pennsylvania I was surrounded every day by images, smells and tastes of farming and food production. Central Pennsylvania's primary economy is farming, everything from commodity crops to specialty vegetable crops for the restaurant trade. I started cooking very early, and knew by the time I was in my teens that I wanted to be a chef. There is something about growing up near the soil surrounded by vegetables and other farm products that inspired me to explore the flavors of food, and specifically how to link the farms to the plates in a way that really expresses the spirit of the produce and the place, that creates a vision for food that's linked to the possibilities of the soil. Tasting the flavors of the region, letting them guide my creativity, knowing what was available when and where, really cemented the idea that the varieties of vegetables and other foods that are indigenous can be used to create meals that truly express the region and the ethos of the land. The French have the concept "terroir" which means a food or wine that is uniquely created by interactions between the soil and culture of a region, something that can only be produced there because of that connection. Here in Pennsylvania I work with farmers, especially Alex Wenger of the Field's Edge Research Farm (who is also profiled in this volume) who grow what they call "landraces" or cultivars of food plants or animals uniquely adapted for a region, with a deep cultural history. Those landraces are better suited for the ecology of the region and have been used and cooked with for generations. They express, like "terroir," what I think of as the "spirit of the place." That's what I try to capture when I cook, I try to use the right food in the right dish so the meal becomes an expression of the region; it's the foundation of my culinary creativity.

After I finished my education at the Culinary Institute of America I worked in some of the most interesting kitchens in the country—at Daniel, Vetri, Le Bec Fin, and Lacroix—and was able to incorporate this vision of the intimate linkage between the land and its food in the restaurant. Because I was working in higher-end restaurants, I could afford to order the best foods from the best farmers—I have had access to really amazing product. Some of the farmers come in to the city with their fresh produce in boxes, just picked that morning, and we could improvise and create with what was perfect that day. It's just so fresh and still linked to the land and the farm that grew it—you can taste that connection.

Reading this chapter about the environmental movement reminds me of how important it is to make sure that the connections between the land, the kitchen work, and the food, and the diner's

FIGURE 3.1 *Chef Steven Eckerd in the kitchen. Photo courtesy of Janet Chrzan.*

FIGURE 3.2 *Steve Eckerd smelling okra blossoms. Photo courtesy of Janet Chrzan.*

FIGURE 3.3 *Chef Steve Eckerd with farmer Alex Wenger. Photo courtesy of Janet Chrzan.*

experience all respect the idea of that food web, the soil, and the place where it's grown. The importance of valuing and taking care of the land so the food is not only healthy for the body but also nurtures the spirit. The idea of the system—that the ecological world is a system we have to respect, it's a food web; we have to respect the connections that make the whole work, that make the food have meaning. When I'm out in Alex's fields, tasting the plants—not just the part we are supposed to eat but the leaves, the flowers, the stems—that plant inspires me, I start thinking about how to respect it as food, how to present it to a diner so that she can also taste the leaves, the stem, the flowers, smell the heat coming off the plants and the dirt in late September. I ask myself how to translate that for the eater. It's the whole experience of the trip to the farm and the tastes of the food that forms my creativity.

A lot of diners are very worried about chemicals in their food. I don't know if they know that much about how the environmental movement has influenced food production, but they do worry about chemicals. I think a lot of them know about the Environmental Working Group's lists of high- and low-chemical foods, the "Dirty Dozen," and the "Clean Fifteen" because they tell us that they don't eat the foods on the "Dirty Dozen." But I wonder if they understand the connectivity of the organic movement, how it's not reduced to a dichotomy of whether X chemical is in food, but how the food is grown, the nurturing of the land, the ability of the land to sustain life, both human and animal, the goals of the organics movement and the environmental movement. I worry they think it's just a checklist item rather than an integrated practice that links soil, plants, and people. It's more than just a certified sticker on the organic produce; it's all the relationships that make that food special to the place and time.

That's why it's important for me to use the local landraces and respect the histories of the food and the people who developed that cultivar and learned to cook with it. If I can use my cooking to explain to a diner how the land and the people and the meal are connected I think I'm starting that dialogue about how and why local and organic food is more than just a thing, is a process that has elements that link to make the whole much greater than its parts.

Organics and the Environmental Movement

Kathleen Delate and Robert Turnbull

From Forests to Farms: A Changing Environment

To understand "Organics and the Environmental Movement," it is necessary to understand the origins of the environmental movement. Like many other stories, it begins in the forest. In *A Forest Journey*, Perlin (1989) examines the role of the forest as a protector and nurturer of soils and a provider of the means for human existence. This treatise serves as an enduring admonishment, warning of the limits of nature, and the consequences of failing to live within those limits.

Approximately fourteen thousand years ago, hunter-gatherer humans began the process leading to the domestication of wheat, barley, rice, millet, pigs, goats, chickens, cattle, and other plants and animals. This effort, combined with a climatic warming following a period of cooling termed the Younger Dryas (c. 11,600 BCE) (Cunliffe 2011), allowed a more sedentary existence, eventually fostering great civilizations, as well as persistent environmental degradation in terms of soil, water, and air. This cause and effect resonates today and forms the foundation of the Environmental Movement.

Plant and animal domestication provided consistent food supplies allowing population growth and clustering. While the time, area, and pace of the transition from hunter-gather to sedentary farmer and to city dweller varied across the globe, this evolution required an ever increasing repurposing of forest, grasslands, and water resources to meet the myriad needs of the dependent populations. Crops and livestock replaced the trees and grasses, and available wood was used to build shelter, cook, and warm the dependent populations. Land clearing of trees and grasses, however, exposed the soil to rain, sun, and wind, degrading its physical structure and fertility, eventually rendering soil resources as unable to meet nutritional demands of crops and livestock. An early strategy employed to address the loss of soil fertility involved changing locations, though its success was dependent on a mobile population and available land. Continued population growth and attendant infrastructure requirements would eventually negate the utility of this strategy in all but a few places and lead to permanent settlements such as Jericho, located on the West Bank region of Israel (c. 10,000 BCE) and Catal Hoyuk (c. 9,500 BCE) in South Central Turkey (Cunliffe 2011).

As populations increased, settlements became cities and commerce ensued, increasing the demand on surrounding resources, especially soil, water, and wood. The advent of copper smelting about 7,500 BCE (Radivojevic et al. 2010) provided a material with more utility than stone with which to fashion, among other things, agricultural tools and weapons. Two to four millennia later, depending on the region, arsenic and then tin was combined with copper, producing bronze about 4,700 BCE (Maniatis et al. 2014), a material far superior than copper in terms of strength and malleability and used to fashion agricultural tools, domestic and sacred objects, building structures, and ship hardware and weapons. By about 3,200 BCE iron replaced bronze in many applications (Asscher 2015). While metallurgy advanced, charcoal derived from wood remained the world's primary source of smelting heat and would remain so until Abraham Darby replaced wood with coked coal in 1709 in England (Fremdling 2000). Growing populations, increasing commerce, and the quest for empire building, requiring ships, soldiers, weapons, and provisions, as well as the various accoutrements of power and security, such as palaces and city walls, continually ran into limitations of natural resources. For example, since trees of sufficient size and character matured slower than human ambition tolerated, the search for forest resources and agricultural lands continually drove conquests of the forests of Southwest Asia, Europe, and the British Isles. The forests were stripped of their cover and wherever possible converted for agricultural production. The wood was used for shipbuilding, weapons, domestic and palatial construction, smelting, currency, and ornamentation. Thus, the maintenance of the empire required access to an ever-increasing exploitable land base and longer, more complex supply channels, often inviting conflict (Perlin 1989).

By the late fifteenth century, much of the world was divided among persistently warring factions seeking some form of control of humans, territory, and resources. This period was also associated with improved oceangoing vessels and navigational experience, allowing seafarers to sail beyond sight of land. While the transatlantic voyages of the Genovese, Christopher Columbus, on behalf of the Spanish Crown between 1492 and 1502 (Johnson 1993) and those of the Venetian, Giovanni Caboto (John Cabot), between 1497 and 1498 on behalf of the English Crown, failed to find a truncated trade route to China (Allen 1992), both demonstrated the efficacy of round trip transatlantic sail crossings and claimed large parts of the Western Hemisphere on behalf of their respective sovereigns. There were, however, profound differences. Cabot stayed only briefly; he left no crewmembers or animals and had little to no contact with indigenous residents (Allen 1992). In contrast, Columbus found people to convert, culture and resources to exploit, and unloaded pigs, horses, cows, wheat, barley, sugar cane, and crewmembers on various Caribbean islands. He also pressed some of the inhabitants into slavery, significantly altered the landscape, and delivered what has been termed "The Columbian Exchange," a lethal combination of measles, tuberculosis, smallpox, influenza, whooping cough, and malaria (Crosby 1967). Over time, these pathogens, often accompanying the tools of warfare, would diminish indigenous human, as well as, animal populations, while universally altering landscapes of the Western Hemisphere and eventually most the world. Over the next three centuries, various explorers were circling the globe charting territory, exploiting human and natural resources, and delivering a suite of diseases to populations with little or no acquired resistance. During this period forestlands of eastern North America would be felled for the same purposes as millennia before with the same devastating ecological results (Perlin 1989). In addition, a plethora of nationalist, religious, racial, and class-based conflicts drove immigration to the Western Hemisphere, increasing agricultural and early industry-driven deforestation.

The certainty of the multiple damaging effects of deforestation was well observed over an extended period of time, and, by 1800, there existed an extensive body of literature dating back to at least the Roman, Pliny the Elder (AD 23–79), outlining strategies to address soil nutrient depletion, soil erosion, and the resulting consequences. These included loss of agricultural productivity as well as silting of harbors and waterways. George Washington (Brooke 1919) and Thomas Jefferson (Betts 1944), whose wealth depended on agriculture, lamented the loss of productive lands and utilized strategies referenced by Pliny, including crop rotations, which required growing crops in a particular order over the course of a number of years in the same growing space to diminish disease pathogens, plant-specific pests and weeds, and replenish desirable plant nutrients. Fallowing—removing land from production for a period of time to allow replenishment of soil nutrients and restructuring of soil—was also advocated (Brooke 1919; Betts 1944).

The Industrial Revolution and the Departure from Nature

In 1827, the Baltimore and Ohio Railroad became the first US railway chartered for transport of passengers and commercial freight, and was but one of the significant changes occurring during the Industrial Revolution (1760–1870). Though coal would replace wood as the dominant fuel where possible, railroad construction in the United States, Europe and its colonies would continue to consume forest resources for rail ties, telegraph poles and accompanying infrastructure (Perlin 1989). In April 1856, the first train crossed the Mississippi River at Rock Island, Illinois, into Davenport, Iowa, and by May 1869, the tracks spanned the nation. Along with transportation, other nascent technologies opened a new era of scientific discovery leading to, among many other things, more efficient manufacturing, large-scale coke-based iron and steel production, and expanding crop production, all of which would have profound effects on the environment (Nriagu 1996).

Increasing the amount of land under agricultural production drove the need for more efficient plant fertilization. One of the first exploratory treatises on methods to improve crop performance was by the German, Carl Sprengel, who published the first articulation of the "Law of the Minimum," examining the role of nitrogen, phosphorus, potassium, sulfur, magnesium, and calcium in plant growth. Publishing in the *Journal of Technical and Economical Chemistry* in 1828, he stated that plant fertilization rates should be based on the minimum amount of a single essential element, and not on the totality of available nutrients. He stated that inorganic materials could provide plant nutrients more effectively than nutrients derived from plant-based humus in soils (van der Ploeg and Kirkham 1999). Sprengel's work formed the basis for the modern fertilizer industry. In addition, in 1844, Dr. John Gorrie, a physician in Appalachicola, Florida, demonstrated the first machine capable of converting water into ice (Gladstone 1998). The eventual combination of increased crop productivity, refrigeration, and rail transportation would allow for more efficient deliveries of greater amounts of food to further destinations. The resulting greater population densities and increased disease epidemics, such as typhus, forced construction of additional infrastructure designed to remove organic wastes from populations (Schultz and McShane 1978).

During the Civil War (1861–5) and continuing until 1916, the Federal government enacted "The Homestead Act of 1862," "The Timber Culture Act" in 1873, the "Desert Land Act" in 1877, the "Timber and Stone Act" in 1878, the "Enlarged Homestead Act" in 1909, and the "Stock-Raising Homestead Act" in 1916 (Anderson et al. 1990) as legislation to encourage land occupation of land west of the Mississippi River. This land was designated for a range of specific purposes, including farming, stock raising, timber harvest, and mineral extraction. During this same time frame, Vermont blacksmith John Deere moved to Grand Detour, Illinois, and introduced the cast steel plow that was patented in 1864, and proved to be particularly useful in plowing untouched grasslands of Illinois, Iowa and beyond. Another key industrialist, Phillip Amour, moved from Milwaukee to Chicago in 1867, joining his brother to build what would eventually become Amour and Company and the Union Stockyards, which ushered in the mass-processing era of the end products of commodity agriculture.

For the first time in the nation's history, the US Census of 1900 noted that despite an increasing population, farming was not the largest employment category. The industrialization of the United States increased labor demand and advances in farming technology, allowed fewer farmers to produce more crops. In addition to John Deere's steel plow that allowed more efficient tillage, Cyrus McCormick's 1834 horse-drawn mechanical grain reaper increased harvest efficiency. In 1911, Fritz Haber and Robert Le Rossignol demonstrated a process during which ammonia was synthesized from air. Along with the escalation of "modern" agriculture across the nation, based on mass markets for agricultural production, farm machinery efficiencies, and improved transportation infrastructure, soils were undergoing increased degradation.

Twentieth-Century Growth of Environmentalism

Theodore Roosevelt is often cited as the politician most influential to the environmental movement; during his eight years in office (1901–9) he not only was a persistent proponent of the conservation of natural resources, but also established the US Forest Service, led by forester Gifford Pinchot. During the next fifty years, the impact of their leadership led to an increasing awareness of the role of the natural world on life in the United States and the impact of human actions on the environment. Much of the environmental movement continued the land conservation mindset, following the works of Henry David Thoreau in New England and John Muir in California. Muir established the Sierra Club in 1892, after his historic perambulations across the country, documenting the beauty of our land. The Dust Bowl, often called one of the worst environmental disasters to ever occur in the United States, spurred the formation of the Soil Conservation Service in 1935. The science of ecology (the study of organisms and their interaction with the environment) was further expanded during this time period, with Arthur Tansley coining the term "ecosystem" (Tansley 1935) and Charles Elton's "food chain" and "food cycle" phrases deployed to describe complex interrelationships among organisms (Elton 1927). The work of environmental ethicist Aldo Leopold, who published *A Sand County Almanac* in 1949, further demonstrated the interactive relationship between humans and nature.

After World War II, the scope of agriculture grew at a rate never before seen. While yields increased, acreage farmed and farm labor decreased. The intensive scale of production was fueled by the burgeoning use of extensive tillage, synthetic fertilizers, and pesticides. However, this intensification came at an environmental cost: soil erosion; nitrates from fertilizer found in drinking water; and the dead zone developing in the Gulf of Mexico. These events did not go unnoticed.

Rachel Carson (1907–1964) came into environmentalism via her love for science and nature (Swaby 2015). Initially known for her poetic writings on wildlife and the oceans, Carson's book, *The Sea around Us* (1951), catapulted her into national attention. She urged the public to understand that "the materials of science are the materials of life itself. Science is the what, the how and the why of everything in our experience" (Carson 1952). When her seminal treatise, *Silent Spring* (Carson 1962), was published, it was heralded as launching the modern environmental movement. In the book, toxic pesticides, and, in particular, the now-banned DDT, were revealed to have long-term cascading effects on the health of wildlife and humans. Carson endured a number of attacks by the chemical industry, but within ten years of the publication of *Silent Spring*, the environmental movement had become socially accepted. April 22, 1970, was the first Earth Day. When the National Environmental Policy Act was passed in December 1970, legislators credited Carson's work with spurring the creation of the EPA, which stands as the most prominent environmental federal agency today.

In some instances, the environmental movement in the United States intersected with anti-war movements, as was the case with pacifists Scott Nearing (1883–1983) and his wife, Helen (1904–1995), who paved the way for the "back-to-the-land" movement by leading a life of rural self-sufficiency in New England starting in the 1930s. Many of the practices advocated by the Nearings in their publication, *Living the Good Life: How to Live Simply and Sanely in a Troubled World* (1954), followed the principles of organic farming: "We are opposed to the theories of a competitive, acquisitive, aggressive, war-making social order, which butchers for food and murders for sport and power. At the same time, and to the utmost extent, we should live as decently, kindly, justly, orderly and efficiently as possible." The 1960s–1980s were a time of transition for the environmental movement as the public's concern for the conservation of natural resources slowly changed to alarm over the threats from toxins in agriculture and industry (Obach 2015). The political turmoil of the 1960s and 1970s instigated the countercultural movement, "including a back-to-the-land element that would ultimately bolster the organic cause" (Lockeretz 2007). Commune farms, food co-ops, and alternative stores and restaurants flourished, with Chez Panisse, Alice Water's paean to local and organic food, opening in 1971 in Berkeley, California. Ecological principles were expanded to include broader social interactions between people and nature, based on "reciprocity and interdependence" (Bookchin 1976; Gershuny 2016), a rallying cry for expanding eco-technologies like organic farming. Environmental organizations proliferated, like the New Alchemy Institute (NAI), also established in 1971, to promote the "creation of ecologically derived human support systems—renewable energy, agriculture, aquaculture, housing and landscapes" (NAI 1989). The 1970s also witnessed the passage of many major environmental legislative acts. In addition to the National Environmental Policy Act, the Clean Water Act, the Clean Air Act, and the Endangered Species Act ushered in a new era of environmental awareness and impact assessments to protect fragile lands and wildlife. John Todd of NAI wrote that "an alternative science is evolving on a world-wide scale" to develop community-based and ecological forces that are "pitted against technological man destroying man and nature" (Todd 1976). The linkage between love of nature and protecting land from destructive forces was on Francis Moore Lappe's mind when she penned *Diet for a Small Planet* (1971), and advocated for a plant-based diet that would have a lower environmental footprint. Her daughter, Anna Lappe, carries a similar message today with her work on the interactions between climate change and fossil fuel–based agriculture (Lappe 2010). The overlap between environmentalists/ecologists and organic agriculturists was evident in the emerging writings of authors like Wendell Berry, who warned that "culture that holds in ignorance or contempt the truths and the mysteries of nature is doomed to failure" (Berry 1976). Other environmental groups, like Ecology Action, in Santa Cruz, California, disseminated information on the power of small-scale agriculture that mimicked natural systems (Jeavons 1971) and set the stage for public monitoring of environmental/agricultural events. Extensive media coverage of the dangers of Alar® growth regulator in 1989 unleashed a firestorm of demands for increased regulation (Neff and Goldman 2005), with environmental groups, such as the Natural Resources Defense Council, working in concert with consumers opposed to chemically treated food.

When much of the ecological and social consequences from energy-intensive commodity agriculture became apparent by the early twentieth century, a new body of scientific literature, eventually termed "agroecology," was offered as an alternative to the status quo. Identifying, examining, articulating, and quantifying the effects of the evolving agronomic system on humans and the environment would be its eventual focus. Wezel et al. (2009) noted that the first articulations of "agroecology" and "ecological agriculture" entered the literature in 1928 (Bensin 1928; Klages 1928) and were used to describe a range of analyses relating to crop production and its environmental effects. Despite the fact that the first book entitled *Agroecology* (Tischler 1965) was only published in 1965, there was already a growing body of research in which agro-ecological principles framed the analytical foundation (Wetzel 2009). However, by combining agricultural production and ecology, the disciplinary bounds of each increased to include myriad social dimensions (Warner 2007).

Again, California was known as the axis for agroecological leaders in the late 1980s. Miguel Altieri offered the first US curriculum in the science of agroecology at the University of California at Berkeley, with the launch of his book, *Agroecology: The Scientific Basis for Sustainable Agriculture* in 1987. Altieri taught that "agroecology is a discipline that provides the basic ecological principles for how to study, design, and manage sustainable agroecosystems that are both productive and natural resource conserving, and that are culturally sensitive, socially just, and economically viable." Stephen Gliessman at University of California, Santa Cruz, also taught agroecology (Gliessman 1996) and helped develop an international Apprenticeship in Ecological Horticulture in association with the University of California, Santa Cruz, student farm, which continues to serve as a model agroecological farm.

The Environmentalist Tapestry of Organic Agriculture

Even with the government-supported impetus for higher productivity with greater inputs—the trading of "soil for oil" (Jackson 1980)—there have always been producers who eschewed the use of synthetic fertilizers and pesticides, either as organic farmers or those simply opposed to embracing this technology. The rise of America's consciousness towards understanding that "when we try to pick out anything by itself, we find it hitched to everything else in the Universe (Muir 1911)" mirrored the rise of the environmental movement and provided the groundwork for an agriculture that recognized the intricate connections between soil, organisms, and plants. The founders of the environmental movement certainly presaged the declining environment associated with the current state of industrial agriculture. Non-point source contamination continues as a major water quality concern in the agricultural lands of the upper Midwest (Humenik et al. 1987) where the pollutant, nitrate-nitrogen, is susceptible to leaching from extensive subsurface tiles draining the land (Magner et al. 2004). The 1992 national water quality inventory (US EPA 1992) found that 72 percent of water quality problems were attributed to agriculture, with plant nutrients the major contaminants of surface water throughout the Midwest (Goolsby et al. 1999). Nitrate contamination from agricultural land in the Midwest, flowing into the Mississippi River, continues to be linked to hypoxia within the Gulf of Mexico (Rabalais et al. 1996). One of the most prominent environmental organization to this day, the Sierra Club, recognizes the importance of promoting a more sustainable agriculture, protective of natural resources: "Agricultural use of land, water, energy and other resources merits high priority but it is also recognized as having among the biggest negative impacts on land and water nationally and globally, so how and where agriculture is carried out is a vital public policy issue" (Sierra Club 2017). Also dubbed as "climate-smart agriculture," organic agriculture strives to meet the criteria for agricultural systems "adapting and building resilience to climate change and reducing and/or removing greenhouse gas emissions" (FAO 2017). It falls upon organic and regenerative agriculturalists to help reverse the damage wrought since the first forest clearings of prehistory to the unabating, massive expansion of monoculture commodity production today.

References

Allen, J. L. 1992. From Cabot to Cartier: The Early Exploration of Eastern North America, 1497–1543. *Annals of the Association of American Geographers* 82: 500–521.

Altieri, M. A. 1987. *Agroecology: The Scientific Basis of Alternative Agriculture*. Boulder, CO: Westview Press.

Anderson, T. L. and Hill, P. J. 1990. The Race for Property Rights. *Journal of Law and Economics* 33: 177–97.

Asscher, Y., Lehmann, G., Rosen, S. A., Weiner, S., and Boaretto, E. 2015. Absolute Dating of the Late Bronze to Iron Age Transition and the Appearance of Philistine Culture in Qubur el-Walaydah, Southern Levant. *Radiocarbon* 57: 77–97.

Bensin, B. M. 1928. *Agroecological Characteristics Description and Classification of the Local Corn Varieties–Chorotypes*. Publisher unknown.

Berry, W. 1976. Where Cities and Farms Come Together. In *Radical Agriculture*, ed. R. Merrill, 14–25. New York: Harper and Row.

Betts, E. M. 1944. *Thomas Jefferson's Garden Book—1766–1824*. Philadelphia: American Philosophical Society.

Bookchin, M. 1976. Radical Agriculture. In *Radical Agriculture*, ed. R. Merrill, 3–13. New York: Harper and Row.

Brooke, W. E. 1919. *The Agricultural Papers of George Washington—1732–1799*. Boston, MA: Richard G. Badger.

Carson, R. 1952. *Acceptance Speech for the 1952 Non-Fiction Award for* The Sea Around Us. New York: National Book Foundation. www.nationalbook.org/nbaacceptspeech_rcarson.html#.WP0OSrGZMU0 (accessed April 10, 2017).

Carson, R. 1962. *Silent Spring*. Boston, MA: Houghton Mifflin.

Crosby, A. 1967. Conquistador y Pestilencia: The First New World Pandemic and the Fall of the Great Indian Empires. *Hispanic American Historical Review* 47: 321–37.

Cunliffe, B. 2011. *Europe between the Oceans 9000 BC–1000 AD*. New Haven: Yale University Press.

Elton, C. 1927. *Animal Ecology*. Chicago: University of Chicago Press.

Food and Agriculture Organization of the UN. *Climate-Smart Agriculture*. Rome: FAO. www.fao.org/climate-smart-agriculture/en/ (accessed April 10, 2017).

Fremdling, R. 2000. Transfer Patterns of British Technology to the Continent: The Case of the Iron Industry. *European Journal of Economic History* 4: 195–222.

Gershuny, G. 2016. *Organic Revolutionary*. Barnet, VT: Joe's Brook Press.

Gladstone, J. 1998. John Gorrie, the Visionary. *ASHRAE Journal* 40: 29–35.

Gliessman, S. R. 1996. *Agroecology: The Ecology of Sustainable Food Systems*. Boca Raton, FL: CRC/Taylor and Francis.

Goolsby, E. A., Battaglin, W. A., Lawrence, G. B., Artz, R. S., Aulenbach, B. T., Hooper, R. P., Keeney, D. R., and Stensland, F. J. 1999. *Flux and Sources of Nutrients in the Mississippi—Atchafalaya River Basin: Topic 3 Report for the Integrated Assessment of Hypoxia in the Gulf of Mexico*. Washington, DC: National Oceanic and Atmospheric Administration. NOAA Coastal Ocean Program Decision Analysis Series No. 17.

Humenik, F. J., Smolen, M. D., and Dressing, S. A. 1987. Pollution from Nonpoint Sources. *Environmental Science & Technology* 21: 737–42.

Jackson, W. 1980. *New Roots for Agriculture*. San Francisco: Friends of the Earth Press.

Jeavons, J. 1971. *How to Grow More Vegetables*. Emeryville, CA: Ten Speed Press.

Johnson, R. 1993. To Conquer and Convert: The Theological Tasks of the Voyages of Columbus. *Soundings: An Interdisciplinary Journal* 76: 17–28.

Klages, K. H. 1928. Crop Ecology and Ecological Crop Geography in the Agronomic Curriculum. *Journal of the American Society of Agronomy* (April): 336–52.

Lappe, A. 2010. *Diet for a Hot Planet: The Climate Crisis at the End of Your Fork and What You Can Do about It*. New York: Bloomsbury.

Lappe, F. M. 1971. *Diet for a Small Planet*. New York: Ballantine Books.

Lockeretz, W. 2007. *Organic Farming: An International History*. Cambridge, MA: CABI International.

Magner, J. A., Paybe, G. A., and Steffen, L. J. 2004. Drainage Effects on Stream Nitrate-N and Hydrology in South-Central Minnesota (USA). *Environmental Monitoring and Assessment* 91: 183–98.

Maniatus, Y., Tsirtsoni, Z., Oberlin, C., Darcque, P., Koukouli-Chryssanthaki, C., Malamidou, D., Siros, T., Miteletsis, M., Papadopoulos, S., and Kromer, B. 2014. New 14C evidence for the Late Neolithic-Early Bronze Age Transition in Southeast Europe. *Open Journal of Archaeometry* 2: 5262.

NAI [New Alchemy Institute]. 1989. Promise Rediscovered: New Alchemy's First Twenty Years. Hatchville, MA: The Green Center. https://newalchemists.files.wordpress.com/2015/01/new-alchemys-first-20-years.pdf (accessed April 10, 2017).

The National Water Quality inventory. The 1992 Report to Congress. Washington, DC: U.S. EPA. https://www.epa.gov/sites/production/files/2015-09/documents/2002_04_08_305b_92report_92summ.pdf (accessed April 10, 2017).

Nearing, S. H. 1954. *Living the Good Life: How to Live Simply and Sanely in a Troubled World*. New York: Schocken Books.

Neff, R. A., and Goldman, L. R. 2005. Regulatory Parallels to Daubert: Stakeholders Influence, "Sound Science," and the Delayed Adoption of Health-Protective Standards. *American Journal of Public Health* 95 (suppl. 1): S81–91.

Nriagu, J. O. 1996. A History of Global Metal Pollution. *Science* 272: 223–4.

Obach, B. K. 2015. *Organic Struggle: The Movement for Sustainable Agriculture in the United States*. Cambridge, MA: MIT Press.

Perlin, J. 1989. *A Forest Journey: Role of Wood in the Development of Civilization*. New York: W. W. Norton.

Pliny the Elder. *The Natural History*. Books 13–18. Translated by John Bostock. www.perseus.tufts.edu/hopper/text?doc=Plin.+Nat.+toc (accessed April 10, 2017).

Rabalais, N. N., Wiseman, W. J., Turner, R. E., Sen Gupta, B. K., and Dortch, Q. 1996. Nutrient Changes in the Mississippi River and System Responses on the Adjacent Continental Shelf. *Estuaries* 19: 386–407.

Radivojevic, M., Rehren, T., Pernicka, E., Sljivan, D., Brauns, M., and Boric, D. 2010. On the Origins of Extractive Metallurgy: New Evidence from Europe. *Journal of Archaeological Science* 37: 2775–87.

Schultz, S., and McShane, C. 1978. To Engineer the Metropolis: Sewers, Sanitation, and City Planning in Late-Nineteenth-Century America. *Journal of American History* 65: 389–411.

Sierra Club. 2017. *Agriculture Policies*. Oakland, CA: Sierra Club. www.sierraclub.org/policy/agriculture/food (accessed April 10, 2017).

Swaby, R. 2015. *Headstrong: 52 Women Who Changed Science—and the World*. New York: Broadway Books.

Tansley, A. G. 1935. The Use and Abuse of Vegetational Terms and Concepts. *Ecology* 16: 284–307.

Tischler, W. 1965. *Agrarökologie*. Jena, Germany: Gustav Fischer Verlag.

Todd, J. 1976. A Modest Proposal: Science for the People. In *Radical Agriculture*, ed. R. Merrill, 259–83. New York: Harper and Row.

van der Ploeg, R. R. and Kirkham, M. B. 1999. On the Origin of the Theory of Mineral Nutrition of Plants and the Law of the Minimum. *Soil Science Society of America Journal* 63: 1055–1062.

Warner, K. D. 2007. The Quality of Sustainability: Agroecological Partnerships and the Geographic Branding of California Winegrapes. *Journal of Rural Studies* 23: 142–55.

Wezel, A., Bellon, S., Dore, T., Francis, C., Vallod, D., and David, C. 2009. Agroecology as a Science, a Movement and a Practice. A Review. *Agronomy for Sustainable Development* 29: 503–15.

PART TWO

Organics in Practice

EDITORS' INTRODUCTION

This section gives the reader a sense of the everyday details of organic crop farming and its potential to positively impact the environment, the farmers, and the financial impacts of these farms—in essence, how the practice of organic farming increases the sustainability of the communities and people it affects, from growers to processors to consumers and eaters. The positive influence of organics is the genesis of its growth as a market sector and its now established place in the food system. As we see from the case study in Denmark, even municipalities have become involved in facilitating the supply chain for organically produced food.

These chapters also emphasize that no matter where in the world, or what crop or animal is being raised, no matter whether large or small scale, organic farming is not easy. There are fewer tools, products, and technologies for farmers to use for the multitude of tasks they face daily; there is little to no government-sponsored price supports, and there are fewer experts for farmers to turn to for advice and information. The challenges faced by smallholder farms in a developing country such as Ethiopia are daunting but necessary to stop the degradation of their land base. In the case of US organic dairy producers, the cost of production continues to increase but retail prices do not. In spite of these hardships, the number of organic farms continue to rise globally, which is a good thing since the demand for organic food is rising as well, and at a rate no one expected. As we read in the previous section, historically, social movements played a role in defining organic food. They continue to do so through the general population's increased awareness of the different values inherent in organic production, even at a large scale. As the number of farmers markets and availability of organic products continue to grow, the increasingly correct public perceptions of the difficult yet worthwhile organic production practices will only serve to spur on further social movements to increase supply and distribution of organic foods.

FIGURE II.1 *Seed spreader for small-scale farming. Photo courtesy of Budd Cohen.*

4

Profile: Alex Wenger, Organic Farmer

Contrary to the lyrics of a popular 1970s song, life on the farm is anything "but kinda laid back." There are always ten different jobs that all need done at once, which is challenging for a one-man work force.

The first few years of my farming career were a painful foray into strong storms, scorching drought, aching back, blistered hands, tired knees, and an exhausted mind. In a typical day I might have jobs like hand planting and updating planting charts to make sure that we never ran out of vegetables to sell, then must turn to a flurry of emails and texts to be sure that what we grew would be sold. Later in the evening I might be on my hands and knees again, pulling weeds from between tiny onion seedlings, or harvesting root vegetables with a shovel. In every way imaginable, small-scale farming is labor-intensive. Growing vegetables organically is even more labor-intensive. We can't rely on powerful chemicals to correct our mistakes. It takes time to grow a ninety-day radish cover-crop that add organic matter and nutrients to the soil. Kaolin clay that protects fruit from insect damage washes off after it rains, and needs to be sprayed on organic fruit more often than insecticides. The chore list is never ending.

Fortunately I have been able to assemble a crew of volunteer workers who love to spend time on our farm, and who have helped me to survive those weeks when I needed extra hands. Chefs love to ride our waterwheel transplanter. My friends who started a raw juice company helped me dig bushels of purple sweet potatoes by hand and plant seeds in the greenhouse in spring and clean in the fall.

It is difficult as a young farmer to get started in farming and to afford the cost of labor and equipment repairs. If I had started my business by paying a working wage to everyone who has helped over the years, the true cost of modern, sustainable food production would have quickly caught up with me. This was also true when prices for tractor parts shot up during the last economic downturn, as fewer and fewer established farmers could afford new machines.

As an organic farm we need to mechanically remove weeds by hand, or cultivate with a tractor so their roots dry in the sun, instead of using a chemical "knock-down" spray to kill weeds, and soil life, at the same time. Insects that chew dozens of Swiss cheese–like holes in our salad green leaves need to be picked off by hand or the plants covered with floating row cover before the insects discover them. Domesticated vegetables have coevolved in an environment where they need careful tending by people. Modern hybrid varieties may pump out tons of fruit per acre, but

FIGURE 4.1 *The Fields Edge Farm. Photo courtesy of Janet Chrzan.*

many modern varieties need to be carefully sheltered and protected in order to do so. Often their root systems are too shallow, and leaves do not grow large, lush, and fast enough to compete with weeds.

Our volunteer crew of artists, musicians, chefs, other farmers, doulas, scientists, and childhood friends also helps with plant breeding projects, as we work to develop the next generation of foods that can be grown organically and better survive disease and insect outbreaks. The goal is to develop crops that are a little more wild, which can survive without such intensive management by people. In one project, we pollinate domesticated apple varieties with pollen from wild apples from Kazakhstan, and grow the seeds of these crosses for four or six years until they make their own fruit. The hope is that some of the offspring of these cross-pollinated fruit will inherit wild survival characteristics from the wild apples, but also keep the traits for productivity that farmers need to be profitable.

Maintaining healthy soil is another major challenge that needs to be solved by thinking in terms of an ecology, not improving individual crop plants. The first few years that I farmed, we sowed oats, clover, and other grains directly on top of the field, following the philosophy of "natural farming," or my interpretation of Masanobu Fukuoka, where clay-wrapped seedballs are spread across the ground. I thought that the plants that I put into the ground would be strong enough to compete with the weeds on their own. Boy, was I wrong, in a big way. A huge weed patch was the only result.

FIGURE 4.2 *Alex Wenger. Photo courtesy of Janet Chrzan.*

One of the best ways that we have found to short-circuit the weed cycle is to build a healthy soil ecology using cover-crops. If we time our planting dates correctly, our cover-crops will smother young weeds before they gained a foothold. As fewer weed seeds contributed to our weed seed bank, our weed pressure fell significantly. Cover-crops also added organic to our soil, which made mechanical cultivation easier. They are also an important source of soil nutrients since we try to use as little manure on our fields as possible, and only organic, antibiotic-free manure at that. For example, buckwheat is a fast-growing, flowering, annual grain crop. In thirty days it can completely cover a bare field with lush, strong stems. When plowed into the soil, the stems begin to decompose, "mobilizing" mineral phosphorus that is otherwise locked in soil. Our buckwheat cover-cropped fields break into a beautiful coffee-color, fine-textured soil.

We plant two-foot-long white daikon, "tillage" radishes to drill into the soil, creating air pockets as they break down, releasing a strong sulfur smell. The first year we grew them in our worst field;

they grew two feet out of the ground instead of deep into the layers of shale! The next year, in that same field, we grew a yellow-blossom sweet clover to build the health of the soil. An unexpected side effect was when the plants erupted in six-foot yellow blossoms, they became a bee magnet. The buzzing of thousands of tiny wings reverberated throughout the field, as bees, wasps, and pollinating insects flew about in huge numbers. This, at a time when pollinating insects have been disappearing elsewhere was remarkable to behold.

Many of these practices of planting cover-crops, cultivating by hand, breeding pest-resistant crops are traditional as well as scientific. I have rediscovered the wisdom of my grandparents along with cutting-edge agronomy research as we designed trial plots to test organic cover-crop rotations aimed at improving soils that were exhausted by the years of poor farming practices. After more than ten years of cover-crops and organic farming, our soil has become moist, dark, with a rich, earthy aroma. In our cover-cropping test plots in 2006 we found six or seven earthworms across a 10-acre stretch of fields. Now I can find hundreds of worms in 1 acre, their breathing tunnels appearing like sculptures after it rains.

Organic farming is beautiful, but also exhausting. Every day can be a test of my endurance. Why go through these struggles and stresses? Because the connection between plants and people is a strong one. There is nothing that can replicate the pure joy when you can hand a refugee a food from their childhood homeland that they haven't eaten for over twenty years. Or when I gave my

FIGURE 4.3 *Alex Wenger with one of the rare plants he grows. Photo courtesy of Janet Chrzan.*

grandmother a special white peach, and watched a smile wash over her face as she remembered her childhood, "This is how peaches tasted in the old days!" I have prepared fresh, flavorful meals for friends, and held near-extinct seeds in my hands. Every time I feel an instant connection to something larger than myself. In my opinion, this is the essence of organic farming, at least how its founders envisioned it. Raising foods without chemical props can be both scientific and traditional. But most importantly new organic farmers need to observe and follow nature's cycles, just as their forefathers did, because farming practices need to work in harmony with all of nature's rhythms.

The Agronomics of Organic Farming

Jacqueline A. Ricotta

Introduction

Much of the time, organic food production is described in terms of what is prohibited: farmers cannot use synthetic pesticides or fertilizers; GMOs are not allowed; sewage sludge is forbidden. But in reality, the agronomics of organic food production is more concerned with what must be done: farmers use composts, manures, and cover crops to improve soils and provide nutrients; pest infestations are prevented through producing healthy plants in suppressive soils, and farm biodiversity is enhanced through crop rotations with cover crops. In fact, the National Organic Program (NOP) describes organic food first and foremost as being "produced by farmers who emphasize the use of renewable resources and the conservation of soil and water to enhance environmental quality for future generations" (USDA National Organic Program 2007). Only then, after emphasizing the positives does the NOP state: "Organic food is produced without using most conventional pesticides; fertilizers made with synthetic ingredients or sewage sludge; bioengineering; or ionizing radiation." Based on both the government standards set in place for certification and the required practices implemented on organic farms, this chapter presents the tools and techniques for production of organic crops as well as discuss some of the theories that seek to go beyond organic.

The National Organic Standards

The National Organic Standards (NOS) are "the complete set of rules that govern organic crop and livestock production and processing" and they "determine what practices and substances are legally required and prohibited to obtain and maintain certification" (Coleman 2012). The NOS also set the standards for product labeling and use of the organic seal. Authorized by the 1990 Farm Bill, the Organic Foods Production Act took twelve years to develop and define the standards, and in 2002 the NOP was unveiled. It is administered by the Agricultural Marketing Service, an arm of the USDA (www.ecfr.gov/cgi-bin/text-idx?c=ecfr&sid=1577ac541828f41863e80e1b1b308198 &rgn=div5&view=text&node=7:3.1.1.9.32&idno=7). Codified in the federal register, the NOS is a long document, but only section 205 applies to producers and processors. It is divided into the following subparts:

Subpart A: Definitions—A glossary of all terms used in organic production

Subpart B: Applicability—Who and what should be certified? An operation "whose gross agricultural income from organic sales totals $5,000 or less annually is exempt from certification."

Subpart C: Production and Handling Requirements—Establishes the three- year time period prior to certification that a farm cannot use any prohibited substances. This section also establishes the "Organic System Plan," which a farmer or handler must submit yearly in order to become and maintain certification. This detailed document (usually a form filled out by the farmer and updated annually) describes everything a farming operation does (from tillage and planting to pest management to harvest, storage, and sales) and what substances will be used. Rules for how and when manures and composts can be used are included in this section. One of the most important inclusions in the organic system plan is how and where records will be kept. In the case of farms that also produce nonorganic products, an explanation of practices and barriers to prevent co-mingling is necessary.

Subpart D: Labeling—Since the NOS came about, the term "organic" has become a legal term. Product labeling is in four categories: *100 percent organic* (all ingredients are certified organic); *Organic*—95 percent or more by weight is certified organic; *Made with organic ingredients*—70–95 percent of the ingredients must be produced according to the NOS; if less than 70 percent of the ingredient are organic, the label can only list the specific ingredient as being organic. Prohibited practices are listed in this section, including the use of genetically modified organisms, sewage sludge, ionizing radiation, and any substances not on the National List of Allowed and Prohibited Substances (see subpart G).

Subpart E: The requirements for certification are explained in this section from the application to the organic system plan, the yearly on-site inspection, and details of recordkeeping.

Subpart F: Accreditation (by the USDA) of Certifying Agents.

Subpart G: The Administrative National List—A comprehensive list of the allowed and prohibited substances for crop and animal production, processing, handling, and packaging.

Any organic farm attempting to become certified should contact an accredited certifying agency (there are approximately eighty worldwide and forty-eight in the United States; the USDA provides a listing to help choose the appropriate certifier: www.ams.usda.gov/services/organic-certification/ certifying-agents) that will provide the information necessary to begin the process as well as have many informational documents to explain the rules and regulations of the NOS.

Managing Soil Health

Whether or not a farm is certified, the most important agronomic practice for the production of high yielding crops of nutritionally dense organic food is the building and promoting of healthy soil, with the key to this being the management of soil organic matter. Soil is defined as the portion of the earth's crust where plants are anchored, and is composed of approximately 45 percent minerals (a mixture of sand, silt or clay); 25 percent air space; 25 percent water; and the remainder organic matter. Organic matter is composed of living organisms (soil livestock); nonliving material such as microbial metabolites, plant litter, decomposing plant and animals; and humus-like substances—an unidentifiable material that is 60–80 percent of the organic matter fraction. And although just a small fraction of the soil profile, organic matter affects almost everything about the soil including improvement of soil structure and water and nutrient retention. Loss of soil organic matter occurs due to erosion, tillage, high temperatures and sun exposure, over grazing, low soil moisture, and application of only inorganic fertilizers. One of the most important roles of soil organic matter is providing a food source for the soil livestock. Each type of organism has a unique role in the soil biome. Through their burrowing movement, earthworms "till" the soil and leave their nutritionally rich casts (feces). Worm castings have been found to be such a particularly rich soil amendment that there are now a number of companies commercially producing them for sale as soil amendments. Soil arthropods such as springtails, sowbugs, and millipedes cut and shred large particles of plant and animal residues; their waste is also rich in nutrients. Bacteria populations in soil are enormous; it is estimated that there are millions of bacteria in a teaspoon of soil. Their job is to release plant nutrients such as nitrogen, phosphorus from organic matter and assist in mineralization, the process through which organic nutrients are converted to inorganic and thus made accessible to plants. Rhizobia bacteria are particularly important since they form a symbiotic relationship with the roots of legumes that enable the plants to fix atmospheric nitrogen and thus provide the basis for the inclusion of legumes as cover crops.

There are numerous types of soil fungi; some help breakdown organic matter while others play a role in decomposition and plant disease prevention. A number of beneficial fungi have been isolated from soil and are now commercial products. For example, Rootshield (Bioworks, Inc. 2014) is a biofungicide composed of the *Trichoderma* fungus; many growers consistently add this to their transplants to protect them from soil pathogens. Other fungi will trap nematodes in their mycelia. *Mycorrhizae* are soil fungi that form a symbiotic relationship with plant roots, and acting like additional root hairs, enable the plant to absorb more water and nutrients. Various species of this organism have been isolated and are sold as soil additives since they provide benefits and enhance growth. Other less important soil livestock include algae, usually found in the upper 0.5 inch of the soil profile; protozoa, which eat bacteria and speed up the nitrogen cycle; and nematodes, the microscopic roundworms that eat plant litter.

Soil quality can be described through physical, chemical, and biological means. Soil tilth, lack of compaction and clods, ability to absorb water, good drainage, overall friability, and a rich earthy smell are components of healthy soil with adequate organic matter and are generally evaluated in the field. Chemically, soil characteristics are measured by soil tests. Farmers send soil samples to their state's land grant university or private soil-testing services, which for a nominal fee will analyze the sample and return the results. Measurements of pH (which should be in the neutral range of

7.0), percentage of organic matter (generally 1–5 percent; gains in organic matter are considered positive) and cation exchange capacity are common measures of soil quality. Biological indicators of soil health include earthworms and microbes, the soil livestock that are critical to breaking down organic matter and providing available nutrients to plants. While measuring worm populations can be done, the number of soil microorganisms is much more difficult and time consuming to quantify.

Nutrient management for organic production

In addition to improving overall soil quality and health, the addition of organic matter provides the nutrients plants require for optimal growth and yields. A mantra of organic farmers is: "Feed the soil, which will in turn feed the plants." Soil test results will indicate if the soil pH is in the optimal range; if too high (alkaline), the recommended soil amendment is sulfur, which in the form elemental sulfur or iron sulfate are allowed in organic systems. If the pH is too low (acidic), lime in the form of dolomitic or calcitic limestone is recommended for an organic farm. All of these materials should be applied well in advance of planting to allow for the changes in pH to occur. Plants grown in soil with the correct pH will be able to absorb nutrients more effectively.

Plant nutrients are divided into macronutrients (needed in larger amounts) (nitrogen, phosphorus, potassium, sulfur, calcium, magnesium) and micronutrients (used in smaller amounts) (chlorine, zinc, manganese, iron), but most farmers worry most about N, P and K (nitrogen, phosphorus, and potassium). A soil test will indicate the relative amount that a farmer's land will need for a given crop; in general, vegetable crops require 75–200 pounds of nitrogen per year. If a soil test is not done, farmers can do a rough calculation of nitrogen credits available from the previous year. Organic farmers obtain these nutrients from a variety of sources, but most commonly used are manure, composts, and cover crops (all considered organic matter) while conventional farms use synthetic fertilizers that have been manufactured at a high energy cost.

The NOP rules for use of raw animal manures are strict to prevent contamination; it must be composted unless (1) applied to land for a crop that will not be consumed (e.g., a cover crop); or (2) incorporated into the soil not less than 120 days prior to the harvest of a product whose edible portion has direct contact with the soil surface or soil particles; or (3) incorporated into the soil not less than 90 days prior to the harvest of a product whose edible portion does not have direct contact with the soil surface or soil particles (NOP section 205.203; see Electronic Code of Federal Regulations, n.d.).

Rules for composting and use of compost are also strict; composted materials must be produced through a regulated and recorded process. The carbon: nitrogen ratio of starting materials should be between 25:1 and 40:1; and the temperature of the pile maintained and recorded for a prescribed number of days depending on the composting system being used. The pile must be aerated as well (NOP section 205.203; see Electronic Code of Federal Regulations, n.d.). The amount of nitrogen in manure (and subsequent compost) depends on the source (cow, horse, poultry) but is generally very low, meaning many tons/acre will need to be applied.

Cover crops

Cover crops play an essential role in soil management and fertility on organic farms. They provide many benefits, including decreasing fertilizer costs, reducing the need for herbicides, preventing soil erosion, conserving soil moisture, and improving overall soil health (Clark 2010). The number of different cover crops and cover-crop "cocktails" (mixtures of crops with different purposes) has grown over the past ten years. Cover crops can be divided into three broad categories: Non-legumes such as annual rye grass, sorghum-sudan grass hybrids, barley, oats, and buckwheat; legumes (which fix atmospheric nitrogen and thus add nitrogen to the soil) such as clovers and vetches; and the *Brassicas* such as mustards, rape seed, and tillage radishes. Choosing the appropriate cover crop for a farm depends on the purpose of the crop (adding organic matter, adding nitrogen, erosion protection, etc.) and the time of year the cover crop will be planted. There are a number of cover-crop decision tools online that allow the farmer to plug in their needs and the program instantly recommends a crop and planting rate (e.g., http://covercrops.cals.cornell.edu/decision-tool.php or http://mccc.msu.edu/selector-tool/).

Cover crops must be managed once planted. In some cases, they are tilled into the soil while still green and growing; these are called "green manures." Some crops are harvested then mown down and baled for hay while others like tillage radishes can be allowed to dissipate in the soil to alleviate compaction. A no-till organic farmer can use a "roller crimper" implement.

Weed control in organic cropping systems

Of the numerous pests that occur in crop production, it is weed control that is generally the most problematic for organic farmers. Weeds compete with crops for light, water and nutrients and decrease or even prevent yields if left to outgrow the crop. Weed identification is an important skill for organic growers as understanding a weed's biology and life cycle can be important to its control (Dupont and Ryan 2012).

Weed management in organic systems is a long-term process that integrates many strategies and tools on an annual basis to deplete the weed seed bank (the population of viable seeds residing in the soil) over time. Growers should do everything possible to ensure that weeds do not go to seed. In the short term, it is most important, if possible, to stay on top of weed pressure from the beginning of the season. Once the weed population grows in size and density, it often becomes a losing battle to control (Schoenbeck 2013). The most important thing an organic grower can do is to plan for weeds. Often it is the crop rotations, tillage, and seeding that are planned for, and when weeds show up (as they inevitably do) the farmer is reactive rather than proactive. Direct seeded crops have a different timing than transplants, and the use of black plastic mulch negates most in-row weeds.

Crop rotation with weed control in mind is important. Crops compete differently with weeds; for example, slow germinating crops like carrots are weak competitors while green beans or corn germinate and emerge rapidly and are easier to cultivate by hand or machine. Incorporating cover crops into a rotation can contribute to a weed control strategy (Grubinger 1997; Mohler 2009). Buckwheat is often used between crops since a good stand will smother out any weeds and prevent them from germinating and going to seed. Some cover crops such as rye and barley release allelochemicals (a chemical produced by the plant to defend itself) into the soil which are inhibitory to certain weeds. Cover crops can be matted down using a roller-crimper to provide a dense covering of the soil; this technique is used for large transplants such as pumpkins or if a farmer has a no-till planter to plant into the rye residue. A living mulch is a cover crop that is planted

FIGURE 4.4 *Tractor-pulled plastic layer. Photo courtesy of Jacqueline A. Ricotta.*

in between the rows of the cash crop to prevent weed germination and growth, but this must be managed so as to not compete with the crop.

Synthetic mulches are frequently used for weed prevention. The standard for commercial vegetable growers is black plastic polyethylene mulch laid with a plastic layer that also buries drip irrigation tape and creates raised beds under the plastic (Figure 4.4). A transplanting implement (such as a water wheel transplanter) punches holes the plastic and transplants are dropped in (Figure 4.5).

The plastic warms the soil and allows for season extension as well as weed control within the rows. Although allowed by the NOS, many organic farmers do not like using a manufactured petroleum-based product such as plastic that must be ripped up (at great effort) and can be difficult to recycle. Organic mulches such as straw may be used within the row or between rows of plastic. In smaller areas such as high tunnels, growers will sometimes cover the entire area with a sheet of plastic and plant into it. A number of farmers are also using landscape fabric between rows of plastic to prevent weeds. This material is porous, reusable, and will allow water to penetrate; it is held down by large staples pushed into the soil.

FIGURE 4.5 *A water wheel transplanter. Photo courtesy of Jacqueline A. Ricotta.*

A temporary solution to weed pressure is mowing or using a weed-whacker. This prevents the weeds from going to seed and can keep them down between rows of plastic or on the edge of fields. Many farmers have had success using propane torches to kill small weeds within rows and there are hand-held as well as tractor-pulled units. Torches produce a small but intensely hot flame that will kill very small weeds (thread stage, which is after the weed has just emerged and its leaves are just beginning to come out, is ideal) but not harm the crop as long as the flame is not held on the plant too long.

Cultivation is often the foundation of weed control on organic farms, especially for row crops such as corn, beans, and small grains. There are many different implements to choose from depending on the particular crop and weed (Grubinger 1997) and larger farms will often have multiple tractors each with a different cultivation implement attached—once the equipment is set up correctly for the crop row spacing, it is easier to leave it on the tractor for the season. Commonly used tractor-pulled implements include field cultivators, finger weeders, basket weeders, brush weeders, torsion weeders, and flex tine harrows (Bowman 2001) (Figure 4.6).

The stale seedbed technique has been around for many years but seems to be having a resurgence in use among organic vegetable growers. To do this, a field is tilled and prepared for

FIGURE 4.6 *A belly-mounted basket weeder. Photo courtesy of Jacqueline A. Ricotta.*

planting; it is then either irrigated or left until it rains. The combination of unearthing weed seed from the soil and water will cause a flush of weeds to appear within a short time. The farmer will then go over the field with a very shallow cultivator (like a flex tine harrow that essentially just drags metal forks over the surface of the soil) or with a flamer that will kill the tiny weeds. The field is then left undisturbed until another rain or irrigation, the weeds are again allowed to flush, and the shallow cultivation is repeated. This entire process can be repeated a third time. What is essentially being done is the weed seed bank in the upper inch or so of soil is being depleted so that by the time the crop is planted, the weed pressure is dramatically decreased.

And finally, no matter how well planned a farm's weed control strategies are, there is the unavoidable hand cultivation. Usually left to interns, this weed control method is time consuming and expensive and often a last resort. There are numerous tools and implements that facilitate weeding and having them readily available is key to effective and timely cultivations. Long handled standard, warren, Korean, grub, collinear, half-moon, diamond, push and stirrup are different styles of hoes. Short handled cultivation tools are effective for small-seeded, closely planted crops such as cilantro. Note that cultivation tools need to be sharpened as often as every day if being used frequently.

Organic insect management

According to the NOS, certified organic growers must use preventative measures first before resorting to the use of an approved substance. The tools of pest management can be broken down into the following categories: cultural controls, which are directed towards the crop and the crop environment; physical/mechanical controls, directed towards the pest; biological controls, which is the use of natural enemies; and finally, there are many insecticides (microbial, natural, and botanical) that can be used for organic insect control.

Cultural methods include having a crop rotation scheme that prevents crops from the same family from being planted near each other for three to five years. This can be difficult to do when farming on a small scale on limited land. Choosing resistant cultivars or varieties that are less desirable to insects can be helpful but growers may have to sacrifice horticultural characteristics (size, flavor, color) for management of a persistent insect pest. Adjustment of a crop's traditional planting date is often used to avoid the injurious stage of an insect pest's life cycle but this technique is sometimes limited by weather and market. Use of season extension technologies such as low and high tunnels may alleviate these limitations (Linker et al. 2009) (Figure 4.7).

Managing soil organic matter and fertility is often a key to limiting pest infestations as it provides optimal crop growth and the ability to withstand pest infestation. Keeping plants watered adequately (1.0–1.5 inches per week is recommended) will also maximize their healthy growth and encourage naturally occurring insect pathogens. Depending on the insect, overhead irrigation will physically knock an insect off a plant. Removal of infested plants and produce will help keep insect pest populations down, and close inspection of any purchased transplants is critical to ensuring that new pests are not introduced to the farm.

Trap cropping entails planting a crop that is preferential to the insect so that the cash crop will have a lower infestation. Many times the trap crop is planted in a perimeter around the cash crop and the insect pest will be drawn to this area. The pest can be controlled via allowed insecticides or the destruction of the trap crop or the farmer may find the infestation on the cash crop acceptable without any pest management.

Physical/mechanical methods attempt to keep the pest away from the crop. There are a number of different types of traps available and can be used for insect population monitoring or destruction. Pheromone traps attract male insects causing mating disruption, and are used very successfully in orchards against insects such as Codling moth. Pheromone traps used in greater quantities can be used for mass trapping, which collects large numbers of insects and thus decreases their population. Yellow sticky card traps are frequently used in greenhouse. Other physical/ mechanical methods include hand destruction—actually picking the large insects off the plant (common with tomato horn worms); using a net to collect the adult cabbage worm moths; or taking off or wiping off leaves that have insect eggs on them (Linker et al. 2009).

Biological controls, which is the use of a natural enemy of an insect pest, have become much more extensive and effective in the past few years. Parasitoids and predators are considered beneficial insects that control insect pests while pathogens such as bacteria, fungi, viruses are components of microbial insecticides, often known as biocontrols. Parasitoids are insect parasites (most often tiny wasps or flies) that lay their eggs inside of the pest at any stage of their life cycle (egg, larvae, adult). The eggs hatch, larvae begin to grow inside of the insect pest, and eventually

FIGURE 4.7 *A low tunnel next to a high tunnel. Photo courtesy of Jacqueline A. Ricotta.*

kill the pest, creating a mummy, the entomological term for the blackened, dried remains of an insect. An example of this would be the *Aphidius* wasp being used for the control of aphids in a greenhouse. Predators, on the other hand, eat their prey, and are generally much less species-specific than parasitoids. Both types of insect natural enemies can be purchased from various suppliers and released, or natural populations can be encouraged through farm conservation measures, such as maintenance of hedgerows and field buffer areas, intercropping, cover cropping, or planting of natural areas to attract insects.

Technically not insects, nematodes are soil-dwelling microscopic unsegmented roundworms. Some are plant pests while others are lethal to insects (entomopathogenic).This type has been used as a biological control agent for soil-dwelling pests and can be purchased and applied or natural populations encouraged through decreased tillage and use of cover crops.

Once relegated to the chemically cautious homeowner, the use of biocontrol agents (insect pathogens, naturally occurring products, botanical insecticides) has grown at a rate that has made conventional chemical companies sit up and take notice to the tune of millions of dollars of investments purchasing the small biocontrol companies and investing millions of dollars in product development. The growth of biocontrol usage has come from the conventional agricultural producers seeking to use them.

One of the oldest and most well-known products contains Bt—*Bacillus thuringiensis*, a soil-borne bacterium that when ingested by insects causes death. Other examples include a fungus, *Beauvaria bassiana*, that infects an insect pest when it comes into contact with its spores. Spinosad is a metabolic product of a bacterium *Saccharopolyspora spinosa*.

Kaolin clay, diatomaceous earth, and sulfur (for spider mites) are compounds that are used in organic production systems to prevent insects, as are various insecticidal oils (usually from plant and fish sources) and insecticidal soaps. Botanically derived insecticides such a pyrethrum (from a chrysanthemum species); neem, from the seeds of the neem tree (*Azadiracta indica*); and various citrus oils (limonene) can also be used if preventative measures have not been successful in controlling insect pests (Caldwell et al. 2013).

Plant disease management for organic crops

A plant disease is caused by pathogens such as fungi, bacteria, viruses or nematodes, and can cause severe financial losses for organic farmers. Organic management of plant diseases attempts to prevent one of the conditions to avoid infection, and the NOS states that certified organic growers must use preventative measures first before resorting to the use of an approved substance.

There are several strategies farmers can use to avoid plant diseases but most important is growing healthy crops through proper cultural practices (fertilization, spacing, irrigation, appropriate weed and insect control) including site selection. This entails avoidance of any problem areas such as low spots, shade, or particularly weedy areas. Overhead irrigation can contribute to infection of fungal pathogens, so many growers use drip irrigation. Excessive nitrogen use can result in overly lush plants that are more susceptible to pathogens. Maintaining a rich soil with regular additions of organic matter is also critical as is crop rotation including the use of varied cover crops. Sanitation such as plowing under crop residue and roguing any diseased plants can be helpful, and growers may want to consider power washing the interior of a frequently used greenhouse with an allowed substance. Choosing a resistant plant or cultivar is often advantageous for the grower, but sometimes does not allow them to grow a preferred variety. Grafting of susceptible varieties onto resistant rootstocks is becoming popular among growers for crops such as tomatoes.

Growers should attempt to exclude the introduction of pathogens by only purchasing disease-free transplants, certified seed, cleaning equipment that has been in an infected area, and thorough cleaning of tools and equipment. Trays used to grow transplants should be cleaned and disinfected before reuse, and potting soil should be kept uncontaminated. Seeds can be treated with hot water to avoid introducing seed-borne diseases. There are a number of microbial seed treatments that provide beneficial bacteria that colonize the root system and protect the plant from fungal growth. The product Rootshield (Bioworks, Inc. 2014) is a *Trichoderma* fungus that can be added to potting mix that prevents damping of fungi. Some potting mixes for commercial growers have started including this biologic control. Along these lines, compost tea is often touted as a disease-suppressive spray but the results have been variable and there are significant food safety concerns.

Inorganic materials such as copper and sulfur are approved for use in certified organic systems and are primarily used by orchardists. Natural materials such as bicarbonate of soda (only approved formulations) can be used as well. The product Regalia is an extract of the Giant Knotweed plant, and when applied as a topical spray will initiate the plants inherent defense system, inducing the plant to produce various compounds to fight infection.

Alternative Agricultural Systems and Theories

Although organic crop production has become relatively mainstream in Europe and North America, there are a number of alternative agricultural theories that, in some cases, pre-date organics and are based on different theories and beliefs but promote an environmentally sound methodology of production with the minimum of off-farm inputs.

Biodynamics

Developed from a series of lectures on "The Spiritual Foundations for the Renewal of Agriculture" given by Rudolph Steiner in 1924, "Biodynamic farmers strive to create a diversified, balanced farm ecosystem that generates health and fertility as much as possible from within the farm itself" (Rudolph Steiner Archives; www.rsarchive.org/). It is also considered by some a spiritual or mystical approach to organic agriculture. Steiner was born in 1861 in Yugoslavia but grew up in Austria. His training in Vienna led him to a career of teaching, writing and giving lectures. He was very impressed with the wisdom of successful peasant farmers, and introduced the concept of astronomy (sun, moon, stars, and planets) affecting the growth of plants.

The basic agronomic principles of biodynamics are similar to those practiced in organics. The farm is considered to be a self-contained entity with its own individuality. There is an emphasis on the integration of crops and livestock with the recycling of livestock waste to help the soil. Plant life is closely related to the life of the soil, and the more vital the food produced from rich soil full of life, the more that food will contribute to the eater's health and vitality.

There are nine different biodynamic preparations that are used in homeopathic quantities to stimulate plant life. The two most well-known and oft-used are horn manure and horn silica. Manure is packed into a cow horn and, on the autumnal equinox, buried in the soil and left until the vernal equinox. Horn silica is powdered quartz that is placed in a cow horn and buried on the vernal equinox and unearthed on the autumnal equinox. The quartz is then placed in glass jars in a sunny location to absorb the sun's energy over the winter before using in the spring. Both preps should be mixed with water and stirred in a vortex for an extensive amount of time (up to an hour) to create a "dynamization," which is sprayed on the crops immediately after mixing. Other preps (yarrow, chamomile blossoms, stinging nettle, oak bark, dandelion flowers, valerian flowers, and horsetail) are also mixed as dynamizations and then sprayed on compost.

Planting of crops in biodynamic farming follows the calendar "Stella Natura" (www.stellanatura. com/use.html), which has been a formal publication since 1978 and follows the work of Maria Thun. Stella Natura takes into account the positions of the twelve constellations of the zodiac, the orbits of the sun, moon, and earth, and the elements of fire, earth, air, and water to recommend when certain crops should be planted. For example, root vegetables like beets or carrots are sown (and will have enhanced growth) when the moons passes in front of a constellation representing an earth element. Farms using biodynamic practices can be certified through the Demeter Association (www.demeter-usa.org/about-demeter/demeter-history.asp), one of the oldest certifying agencies worldwide.

Permaculture and natural farming

Permaculture, from the conjunction of "permanent agriculture" is not an agricultural production system but instead a land use and community planning philosophy proposed by Bill Mollison and David Holmgren in the late 1970s (Mollison 1988). Designing ecological habitats in conjunction with food production systems, the concept includes the integration of plants, animals, soils, and human dwellings into productive, stable entities. Modeled after nature, relationships among the stakeholders rather than individuality is stressed. The ethics of permaculture include care for the earth and of people, setting limits to consumption, and the values of all living organisms.

When used in the design of a farm, homestead, or community, permaculture emphasizes synergy between site components, organic and sustainable production, and a holistic, diverse ecosystem that is productive. In the book *Restoration Agriculture*, Mark Shepard (2013) reinterprets permaculture and focuses on the production of perennial crops and the concept of a "food forest"—a woodland ecosystem with edible crops. Permaculture advocates are beginning to plan food forests in urban areas such as the one started by the Urban Ecology Center in Washington Park in Milwaukee, Wisconsin.

Philosophically, Mollison and Holmgren were strongly influenced by the work of Masanobu Fukuoka, author of the *One Straw Revolution* (1978) and the proponent of "do nothing farming," which is also known as natural farming. Fukuoka was born and raised on the island of Shikoku in Japan, and trained as a plant pathologist. Giving up his career to return to the farm, he writes about the need to work with nature to provide healthy, wholesome food without harming the very land that provides the harvest. Fukuoka also advocates small-scale farms and the rejection of the industrial model of farming: "If natural farming were practiced, a farmer would also have time for leisure and social activities within the village community. I think this is the most direct path toward making this country a happy pleasant land" (Fukuoka 1978).

Summary

Organic crops are produced on thousands of farms worldwide, and every farmer uses their own experience, knowledge, beliefs, and judgement to result in a profitable harvest and clean, healthy food. The agronomic practices discussed in this chapter are a summary of the many facets of production as recommended by the USDA and scientists from agricultural research institutions. It is up to each individual farmer to create the system that works best for their particular climate, soils, and markets and provides the necessary yields to maintain profitability while enhancing the environment of their farm and community.

References

Baldwin, K. R. 2006. *Soil Fertility on Organic Farms*. Raleigh: North Carolina State University Cooperative Extension Service.

Bioworks, Inc. 2014. www.bioworksinc.com/products/shared/rootshield.pdf

Bowman, G., ed. 2001. *Steel in the Field: A Farmer's Guide to Weed Management Tools*. www.sare.org/Learning-Center/Books/Steel-in-the-Field

Caldwell, B., Sideman, E., Seaman, A., Shelton, A., and Smart, C. 2013. *Resource Guide for Organic Insect and Disease Management*. Cornell University Agricultural Extension Service. http://web.pppmb.cals.cornell.edu/resourceguide/

Clark, A., ed. 2010. *Managing Cover Crops Profitably*, 3rd edn. Sustainable Agriculture Research and Education.

Coleman, P. 2012. Guide for Organic Crop Producers. National Center for Appropriate Technology. www.ams.usda.gov/sites/default/files/media/GuideForOrganicCropProducers.pdf

Cover Crop Decision Tool. http://covercrops.cals.cornell.edu/decision-tool.php

Dupont, T. I. 2012. *Soil Quality*. Pennsylvania State Cooperative Extension.

Dupont, T. I, and Gugino, B. 2012. *Ecological Disease Management*. Pennsylvania State University Extension. http://extension.psu.edu/business/start-farming/vegetables/factsheets/ecological-disease-management/extension_publication_file

Dupont, T. I., and Ryan, M. 2012. *Creating a Weed Management Plan for Your Organic Farm*. Pennsylvania State University Extension. http://extension.psu.edu/business/start-farming/vegetables/factsheets/creating-a-weed-management-plan-for-your-organic-farm

Fukuoka, M. 1978. *The One-Straw Revolution*. New York Review of Books.

Gold, M. V. 2007. Organic Production/Organic Food: Information Access Tools. Alternative Farming Systems Information Center; National Agricultural Library; US Department of Agriculture.

Grubinger, V. 1997. 10 Steps toward Organic Weed Control. *American Vegetable Grower* 46: 22–4.

Grubinger, V. 2008. *Nutrient Management on Organic Vegetable Farms*. University of Vermont Cooperative Extension Service.

Koike, S. T., Gaskell, M., Fouche, C., Smith, R. and Mitchell, J. Plant Disease Management for Organic Crops. http://anrcatalog.ucanr.edu/pdf/7252.pdf

Linker, H. M., Orr, D. B., and Barbercheck, M. E. 2009. *Insect Management on Organic Farms*. Raleigh, NC: North Carolina Cooperative Extension Service.

Mohler, C. 2009. Crop Rotation on Organic Farms. Naturel Resources and Engineering Service. www.sare.org/Learning-Center/Books/Crop-Rotation-on-Organic-Farms

Mollison, B. C. 1988. *Permaculture: A Designer's Manual*. Tyalgum, Australia: Tagari.

Electronic Code of Federal Regulations. *Part 205: National Organic Program*. National Government Publishing Office. www.ecfr.gov/cgi-bin/text-idx?c=ecfr&sid=1577ac541828f41863e80e1b1b308198&rgn=div5&view=text&node=7:3.1.1.9.32&idno=7

Sanchez, E. S. 2009. *Using Organic Nutrient Sources*. Pennsylvania State Cooperative Extension.

Schoenbeck, M. 2013. Twelve Steps toward Ecological Weed Management in Organic Vegetables. *eXtension.org*. http://articles.extension.org/pages/18539/twelve-steps-toward-ecological-weed-management-in-organic-vegetables

Shepard, M. 2013. *Restoration Agriculture*. Greeley, CO: Acres USA.

USDA National Organic Program 2007. www.nal.usda.gov/afsic/organic-productionorganic-food-information-access-tools

Wyenandt, A., ed. 2016. Mid-Atlantic Commercial Vegetable Recommendations. http://extension.psu.edu/publications/agrs-028

www.federalregister.gov/documents/2014/09/30/2014-23135/national-organic-program-nop-amendments-to-the-national-list-of-allowed-and-prohibited-substances

www.sare.org/content/download/29730/413972/Steel_in_the_Field.pdf

5

Profile: Alex Wenger, Organic Farmer

Every visit to our friend's goat dairy and cheese room is a cultural experience. Goats, kids, children, and adults all mingle. Milking, cheese-making, goat caramel cooking, retail sales, and the occasional tour bus keep the family busy. On-site cheese-tastings, complete with a bubbling kombucha pairing greet guests, and as they step into the small store which is full of carefully packaged products that range from floral, moisturizing, goat milk soap, to the amazingly rich creamy goat milk caramel. Amos, cheese-maker and marketer extraordinaire, is not your average farmer. He has chosen to grow differently, and take on the risk of developing and marketing new food products himself, as he explained to me one day when I went to visit his cheese-making operation.

Rolling through the lush, jungle-like cornfields of Lancaster, Pennsylvania, herds of dairy cows make it feel like one has entered an agricultural paradise, a land that time has forgotten. Nothing could be further from the truth. Most farms in Lancaster are tied into one of the largest commodity markets in the world. Many dairy farmers, even some organic farmers, have trouble making enough profit to survive when milk prices dip suddenly with "shocks" influenced by Wall Street investors, or production in on other continents. Many farmers feel pressure to raise commodity foods, which require huge volumes of product to turn a profit. Former secretary of agriculture Earl Butz once said that farmers have the choice to "get big or get out," of farming.

Farmers can also choose to "go niche," and serve specialty markets instead. Some local farmers like Amos have turned to cheese making to turn their milk into high value-added dairy products like cheese and yogurt, to capture more of the food dollar. Amos has taken this philosophy and run with it. He depends on maintaining a strong niche for his family's survival. Goat milk itself is not a commodity. Each gallon sells for several times the price of organic cow milk. Amos is not certified organic because his customers trust his pledge to keep chemicals off his farm, and out of their food. They can visit the farm in person to see Amos's farming practices in action as well.

If Amos were producing a homogeneous commodity, he might have fewer jobs on his to-do list every day. To ensure income for his family he has to carefully manage all aspects of his business: from the breeding of the goats, to sourcing their chemical-free feed, to quickly switching cheese recipes if the flavor of his milk changes. And then he makes phone calls, prints brochures, directs cheese tastings, and markets his products. These strategies are more complicated than sending his milk to a factory to be blended with milk from dozens of other farms, but, so far, these strategies have ensured his farm's survival.

FIGURE 5.1 *Dairy goat. Photo courtesy of Janet Chrzan.*

Organic dairy once offered farmers a niche marketing strategy to recoup some value for their milk. Nevertheless, Amos avoided it completely since organic milk has become a commodity. Another farmer friend of ours tried turning to organic, pastured dairy as a strategy to grow his small farm. His milk quality was exceptional, and the co-op that he sold to shipped it to California for cheese-making purposes. Even so, he couldn't make enough money with his dairy with twenty-five cows, and he had to choose to either scale up his herd and workload to over forty milk cows, or get out of the business. He chose to get out of the business, and work off the farm. The organic milk co-ops that once helped small farmers are growing larger to cater to an industrial food supply chain whatever the product. When farmers enter an industrial market, they all face similar challenges.

We work with a similar marketing and production dynamic to Amos as we developed and marketed a specialty heirloom product as part of our farm's offering. Its name in Italian, Spin Rosso della Valsugana, roughly translates as "red-spined corn from the Sugana Valley," which is a valley in the Alps that stretches across northeastern Italy. I tracked down seeds of this near-extinct corn because I heard that it made "the very best polenta." The Spin Rosso corn is a landrace, which means that it has wild genetics. Unlike modern corn, each ear of this variety grows out at a different height, the plants themselves grow from six to ten feet tall, which is a contrast to the soldier-like hybrid corn that is grown in much of Pennsylvania, adapted to factory farming with large equipment.

The first year that we grew this corn, all but 10 percent of the stalks fell over in the wind. By walking through the field and saving seeds of the best ears each year, we have slowly adapted this variety to our farm, just as Amos has carefully bred his goat herd. I had to develop a system for harvest and processing for this corn, and added roasting as a traditional value-added process. I worked with a local stone mill to process large volumes as we scaled up and then I developed packaging and marketing to sell the floral, nutty cornmeal. Several chefs have designed dishes for their restaurants in Lancaster, Philly, and New York that feature our corn, which I can sell for

FIGURE 5.2 *Alex Wenger and Steve Eckerd in the cornfield. Photo courtesy of Janet Chrzan.*

several times the price of commodity grain corn. Just like Amos and his dairy niche, we've created our own niche.

Organic dairy was once cutting-edge, but now many dairy farmers are recognizing that milk is market-controlled and industrially handled. Those who invest extra time and labor in creating a value-added product like cheese, yogurt, or ice cream will be able to compete despite a commodity market, if they do a good job of building trust and recognition among consumers. Getting started with these businesses is challenging, just as challenging as it was to improve the genetics of our corn, learn how to create corn meal, and market the finished product myself. But both Amos and I can receive a far higher price for the food that we grow, speaking to the power of marketing food that is good for people, and the planet, relying on our community of committed customers.

The Organic Dairy Industry: How Markets, Standards, and Technology Lead to Changes on the Farm and Throughout the Supply Chain

Adam Diamond

An early focus on niche, local markets and an orientation toward high-quality, unprocessed food in the organic sector is giving way to mass-produced organic food sold through mainstream channels. The organic dairy sector is a particularly interesting case because, in comparison to its conventional counterpart, it is more linked to branding, public relations, and advanced processing technology, even as significant motivators for organic milk consumption and production include support for local economies, slowing down production and consuming more natural, that is, unadulterated food products (Hill and Lynchahaun 2002). The marketing intensive character of the chain puts pressure on price premiums and leads to geographically dispersed yet economically centralized production and marketing networks. It even influences milk processing technology and the taste of the product, in what some would say an inferior direction (Swaminathan 2008; Michaelis 2013), even as farmers are reducing production per cow and moving away from reliance on purchased inputs. This chapter seeks to explain how and why the organic dairy industry is developing the way it is, highlight some of the key controversies and issues going forward, and offer up suggestions for supporting its integrity and expansion.

The Agricultural Treadmill and the Organic Dairy Commodity Chain

The very same forces of globalization, economic concentration, and relentless technological "improvement" that catalyzed the development of organic dairy farming as a way to slow down the relentless march of the agricultural treadmill insinuate themselves into various stages of organic food commodity chains, albeit in different ways than with conventional agricultural production. The agricultural treadmill (Cochrane 1993) addresses the tendency within American agriculture towards greater adoption of technology and increasing productivity, leading to concentration in land ownership. Farmers adopt innovations in the hopes of increasing profitability, yet gains from such innovations are largely dissipated through lower prices brought on by increased supply, or higher land prices when prices are propped up by subsidies. The government aids this process with extension services and research oriented around increasing productivity, but Cochrane (1993) does not believe the enduring dynamic that is the agricultural treadmill can be altered. Cochrane's agricultural treadmill is paralleled by a more general technological treadmill in industrial capitalist societies, wherein over time production becomes more capital intensive (Schnaiberg 1980). Unlike Cochrane, the treadmill of production school within sociology (Schnaiberg 1980; Schnaiberg and Gould 1994; Weinberg et al. 2000; Gould et al. 2008) holds that this tendency can be either slowed or reversed under certain conditions, namely sufficient resistance to the treadmill by citizen-workers who see their well-being damaged by this pervasive phenomenon of developing and deploying new technologies that put people out of work and degrade landscapes and ecosystems. This process clearly applies to the organic certification regime as various environmental and agricultural groups have mobilized to pass organic legislation, expand regulation and enforce regulation, all the while battling economic interests that seek to weaken the standards in the interests of expanding the organic market, which they argue would increase the overall environmental benefit flowing from the standards (Obach 2007).

Within the organic dairy industry there is a constant and fundamental tension between efforts to expand the market, whether through loosened standards, organic livestock concentrated animal feeding operations (CAFOs), extended shelf-life or international exports, on the one hand, and efforts to restrict access to the organic label to those farms most dedicated to organic principles, and to ensure that it remains more of a protective niche for smaller, lower input dairy farms, on the other. The USDA's oversight of the organic dairy industry certification regime produces contradictory effects because of the government's dual role as economic growth promoter and agent of social consensus (Gould et al. 1996; Novek 2003). The USDA's National Organic Program, through its promulgation and enforcement of federal organic standards, accreditation and oversight of organic certifiers, and establishment of international trade agreements for organic products, aims to create new business opportunities through the marketing of third party certified, differentiated products,[1] while also seeking to transform agriculture by stimulating new forms of farm-level production based on lower input usage, nutrient cycling, and rejection of harmful synthetic fertilizers and pesticides. Sometimes the interests of organic farmers and organic food processors/marketers appear to be at odds with each other regarding standards setting and interpretation; farmers want strict standards and premium farm gate prices, while processors and marketers would like to see a plentiful supply of organic food commodities to lower their input costs. If that means the standards are not quite as strict then so be it.

In organic dairying the returns to farmers can increase as the treadmill of production is slowed down. Rather than investing in labor-saving and yield-enhancing technologies to increase profits, organic dairy farmers in the United States have lower production per cow, have smaller farms, and

FIGURE 5.3 *Dairy equipment. Photo courtesy of Adam Diamond.*

use more labor to meet organic standards (Guptill 2009). Following these practices allows them to earn a significantly higher price for their milk, upwards of 100 percent over the conventional milk price. However, this slowing down of the treadmill at the farm level coexists with changes in distribution and marketing patterns within the commodity chain that operate in tension with farm level trends. The presence of geographically dispersed distribution chains, widespread use of pasteurization technology that makes the milk less perishable, and economically concentrated marketing networks point to a more technologically advanced, less competitive distribution and marketing node that fosters power imbalances within the commodity chain.

The commodity chain framework is used to focus attention on the relationships between different actors (Gereffi and Kornzieniwicz 1994; Sturgeon 2009) within the organic dairy industry, how power infuses itself within these relationships, and how the geographic dispersion of production affects these relationships and, in turn, treadmill dynamics. Commodity chain analysis is particularly useful as a technique for making more transparent the power dynamics, goals, and distribution of benefits within industry such as the organic dairy sector that is so laden with symbolism and ethical values or motivations.

FIGURE 5.4 *Organic, pastured Jersey cows. Photo courtesy of Janet Chrzan.*

Organic Certification as Counter to Intensification and Consolidation

Generally, commercial farmers in advanced industrialized countries face a cost price squeeze in which powerful input suppliers and processors capture the lion's share of value added (Mooney 1988; Goodman and Redclift 1991), pressuring farmers to adopt more and more "efficient" technologies to lower their cost of production. However, this drive for efficiency produces significant environmental externalities, that is, damage to natural resources and ecosystems that is not reflected in the price consumers pay, and reduces the overall number of farmers. There is a positive feedback loop as higher productivity increases pressure on surviving farmers to increase their productivity (Cochrane 1993), which only furthers farm consolidation as food has relatively inelastic demand. As production per farm goes up faster than demand inevitably the number of farms goes down. Organic certification runs counter to this pattern as it gives economic value to more ecologically sustainable production methods that often are less "efficient" in terms of yield per acre or person hour of farm labor (Guthman 2009).

The agricultural treadmill has accelerated quite rapidly in the dairy industry over the last four decades, with the number of dairy farms declining by more than 90 percent, from 1.13 million in 1964 (Blayney 2002) to 49,629 in 2012 (National Agricultural Statistics Service 2014). Consolidation at the farm level has been accompanied, and perhaps accelerated, by a parallel concentration in

FIGURE 5.5 *Grain silo. Courtesy of Adam Diamond.*

milk processing that has led to a dramatic reduction in the number of milk processing plants—from fluid milk to cheese and butter, from New York, to Wisconsin and California (Lyson and Geisler 1992; Lyson and Gillespie 1995). According to the National Agricultural Statistics Service, from 1970 to 2015 the number of dairy product–manufacturing plants in the United States declined by 66 percent, from 3,749 to 1,267.[2]

The Role of Marketing Firms as Commodity Chain Drivers in the Organic Dairy Industry

Organic dairying is a new form of production that slows down the treadmill of production at the farm level even as the treadmill of production continues to operate at a fast rate—or even increases in speed—at the distribution and marketing levels. The price premium afforded organic dairy farmers, which represents a combination of scarcity induced rents and unavoidable higher costs of production, is vulnerable to downward pressure when farmers have an asymmetric relationship with their milk buyers.

Organic dairy companies manage commodity chains from farm to lips. They contract with farmers to raise cows and produce milk, with processors to bottle the milk, and with distributors or trucking companies to deliver milk to retail outlets. Completing the chain, the two main firms have national sales and marketing forces that support and develop existing accounts with retailers, reach out to new retail outlets, be they natural food chains, supermarket chains, warehouse stores like Costco, or independent grocery and natural food stores; target customers through social media and sponsorship of public radio; and arrange for farm visits with suppliers to literally connect consumers with the source of their milk. They own very little in the way of distribution or processing facilities. Organic Valley owns three dairy processing plants and one distribution center; it contracts out all other production with ninety-five plants around the country and leases eighteen warehouses. Horizon works with forty-five plants to process and package its products. A vice president with QAI, one of the biggest certifiers of organic food manufacturers, commented on this lack of infrastructure investment.

> You know … My favorite question to people is what do organic manufacturers manufacture?
> And the answer is … labels, because many organic manufacturers don't manufacture. They subcontract it to other places that run it for them, and put their … focus on brokering, marketing and distribution.

In part, this lack of ownership of manufacturing facilities by organic food companies is an artifact of scale. As they are essentially bit players in the $1-trillion-plus US food business, they simply cannot manufacture on the scale necessary to justify owning dedicated manufacturing facilities. However, scale alone fails to explain this dynamic as growth in the organic industry has, if anything, led to more contracting out of production. This same QAI representative described this process as follows:

> The trend seems to be the reverse. When Hain bought Walnut Acres, Walnut Acres had their own manufacturing facility. That was scrapped. It's all made in New Jersey now.

In organic food marketing the money lies in control of the label, not manufacturing. There is little to be gained through ownership of facilities, while potential losses are greater when precious capital is tied up in fixed infrastructure. In this manner firms avoid the problem of sunk costs, that is, capital expenditures that cannot be recovered (Clark and Wrigley 1997), by minimizing investment in fixed capital. Contracting out production provides firms more flexibility to respond to fast changing markets and allows the firms to concentrate on arranging for adequate supply and marketing products.

Each firm constructs its own story in connection with the general attributes of organic to attract consumers. A marketing representative from CROPP/Organic Valley described this process as follows:

> We are a public relations driven company, and that's how I'd describe our outreach to the consumer … One of the things that we do a lot of is we press out stories of our farmers, we press out our commitment and our story about being a life-line to farmers in rural America.

One can see in vivid pictures, text and videos how CROPP/Organic Valley uses farmer stories to reach customers on its website at www.organicvalley.coop/

While the logistics behind the scenes are key to making the marketing effective, it is the marketing that is driving the chain. Farmers sometimes try to get involved with the marketing side, by working with neighboring farmers to jointly process and market their milk together, knowing that this is where a large proportion of value is generated. However, this strategy is difficult to implement given the stiff competition faced by local processors from the national firms with highly trained marketing staff and the time constraints faced by farmers who have more than enough challenges just handling farm management, let alone the complexities of aggregation, processing and marketing of milk. One New York farmer who used to sell to a small organic cheese cooperative and then switched to CROPP/Organic Valley after the cooperative failed to pay even conventional milk prices because of difficulties in marketing the cheese reflected on his experience: "it's all in the marketing. But the problem is—farming is a full-time job."

The power of firms revolves around their de facto ability to restrict access to this coveted label, which functions as a barrier to entry (Guthman 2009). Firms, informed by projections of market demand and evaluations of logistical convenience, have a great deal of influence over whether a particular farmer can get on the milk truck and thus gain access to this lucrative market. Beyond serving as gatekeepers and product marketers, the firms also determine how much producers are paid, the stability of farmer pay prices, how and where it is processed, and the extent of the distribution network.

Milk Marketing Firms Contract with Farmers to Feed the Marketing Treadmill

The two main firms contract with more than 2,000 farms (Horizon—650+, Organic Valley—1,500+) accounting for more than 75 percent of organic dairy farms in the country,[3] who in turn work with dozens of certifiers around the country who operate as gatekeepers under the supervision of the USDA's National Organic Program. As managers of the entire chain the organic milk firms exercise considerable leverage over individual farmers, or even groups of farmers. When sales are slow, conventional dairy farmers transitioning to organic will have a harder time getting on the organic milk truck, while farms currently selling organic milk may have to accept small price cuts. In some cases, farmers have had to accept quotas whereby all supplying farmers agree reduce their production by a small percentage. On occasion, farmers in remote areas far from other organic dairy farms or milk plants have been cut off when sales softened and their contracts expired.

Organic Milk Pricing

In general, organic dairy farmers receive substantial premiums for their milk over what conventional dairy farmers receive. Current pay prices offered by Organic Valley, the largest buyer of organic milk in the country, range from $33.43 to $39.68 per 100 lb (www.farmers.coop/producer-pools/dairy-pool/pay-price/). These prices vary depending on the region of the country and the type of milk, with "Grassmilk" from cows only fed grass and hay earning a substantial premium over milk from

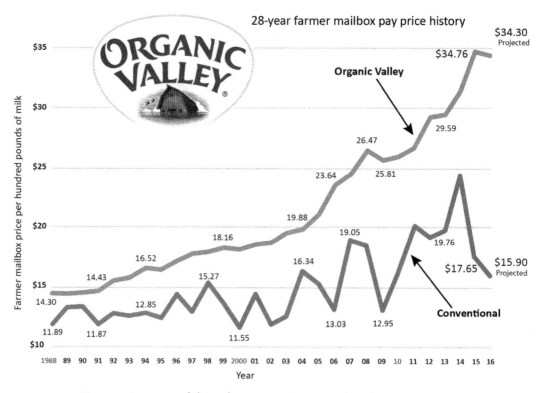

FIGURE 5.6 *Twenty-six years of dairy farmer pay. Reprinted with permission.*

cows that are also fed grain. This compares to the all milk price for June 2016 of $14.80 per 100 lb, down from $17 per l00 lb for June 2015. While Organic Valley is not currently the price leader, with some firms paying slightly more in some regions, it does buy 50 percent of organic milk produced nationally, and Horizon Organic, which buys 25 percent of organic milk produced nationally, has to keep up with Organic Valley to maintain its supply (Su et al. 2013). Overall, the prices for organic milk have been steadier than conventional prices, and on a steady upward trajectory, such that the price premium has increased over time in absolute and relative terms. See Figure 5.6 for a visual demonstration of this stability of organic pay prices for the market leader Organic Valley, compared to conventional milk prices.

This steady pay price for organic milk has been of tremendous benefit to organic dairy farmers, allowing them to better predict their cash flow from month to month and year to year.

Greater price stability is one manifestation of how the organic certification regime has slowed down the dairy treadmill of production whereby highly variable prices in concert with long-term gradual price declines have put pressure on dairy farmers to invest in more efficient milking technologies, feed more grain to cows to increase milk production per cow, and increase herd size to spread equipment costs over a larger production base. However, even with these higher and more stable prices for organic milk, organic dairy farmers and organizations have repeatedly called for higher prices to cover the dramatically higher cost of organic feed (Maltby 2006;

Richardson 2008; Hill 2012; Maltby 2012) which can be twice as expensive as conventional feed. The premium at the farm gate is significantly lower than the premium at the retail level (Figures 5.3 and 5.4). While organic milk retail prices were consistently more than twice those for conventional milk, farm gate prices for the former have ranged from just over 25–100 percent higher than the latter.

On average, the distribution, marketing, and retail components of the organic milk commodity chain account for a greater share of retail milk prices than with conventional milk. In part this derives from the geographically dispersed character of organic milk distribution chains and their attached higher transportation costs. The niche character of the organic market requires larger aggregation and distribution sheds for any given processing plant in comparison to conventional milk. Organic milk usually travels farther from farm to processing plant because organic dairy farms are sprinkled amidst a sea of conventional dairy farms, requiring milk trucks to travel longer distances between farms to pick up a full load to drop off at the bottling plant, to say nothing of the fact that organic farms are smaller on average than conventional farms, requiring more stops to fill up a milk tanker truck with 30,000 lb of milk. And after the milk is bottled it travels farther to stores. Organic standards restrict the scale of farm operations because the requirement to pasture milking cows throughout the lactation cycle is very difficult to meet as herds get above 500–600 cows (Diamond 2013). However, the standards do not restrict the scale of milk supply networks. In fact, it seems that the niche character of the organic market encourages larger scale supply networks, due to the smaller size of individual farms as well as their relative scarcity amidst the dairy landscape.

Processing Technology as Tool for Dispersing Milk Distribution

One technology that has accelerated both the geographic dispersion extent of distribution chains and the concentration at the firm level is ultra-high-temperature (UHT) pasteurization. Instead of using traditional or high-temperature short-time (HTST) pasteurization, which has a seventeen-day shelf-life, the two national organic dairy firms have used UHT pasteurization extensively as a method to extend shelf-life and thus the distribution chain. UHT milk, heated to 280 degrees and injected with steam to prevent scalding, has up to a seventy-day shelf-life.

While conventional dairy companies pioneered the use of the technology for specialty milks such as lactose free or chocolate, Organic Valley pioneered the use of UHT pasteurization for white milk, and this technology has since been much more widely adopted by the organic dairy industry as a way to extend shelf-life, allowing milk to sit longer, either in a warehouse or on the retail shelf. The longer sell-by dates played a crucial role in initial retail acceptance as dairy case managers did not have to worry about rotating the organic section every few days. A marketing representative from Horizon described the rational for UHT pasteurization as follows:

When it was starting out, gosh it's been a long time now, but several years ago, organic milk didn't turn as quickly ... So in the ultra-pasteurized technique you're able to get a little more code date, a little more shelf life, and that gave our customers, our retail customers, the opportunity

FIGURE 5.7 *Dairy section. Courtesy of Adam Diamond.*

to sell through the product before it would spoil … So that was originally the reason that we started offering ultra-pasteurized. Really as a company that's growing we were looking for all different ways that we could offer milk to our customers.

Besides alleviating concerns about production rotation at the retail level, UHT technology also helps minimize product loss and meet changing demand. A sales executive with Organic Valley explained, "Regular [HTST] milk has to be sold very quickly, but if demand slackens then it could have to be dumped, or put on drastic markdown. Likewise, if demand increases there may not be any available, whereas with UHT inventory can be built up as a buffer against these swings."

There is a push–pull relationship between UHT processing and dispersed distribution patterns. The use of UHT processing has accelerated the expansion of the organic milk commodity chain by increasing shelf-life, while in turn its use has been fostered by the geographically dispersed character of the distribution network already in place when it was introduced, which is itself an artifact of the niche character of organic products (Watts et al. 2005). When UHT technology was introduced, Horizon and CROPP/Organic Valley were already building national brands, and the distribution networks to support them. Adoption of UHT technology greatly facilitated this process. Horizon introduced UHT technology in 1999, and their sales have rapidly expanded from $49 million in 1998 to $644 million in 2014 (WhiteWave 2014). Likewise, CROPP/Organic Valley introduced UHT milk in 1998, and their sales have similarly exploded, going from $28 million in 1997 to $1.04 billion in 2015 (CROPP/Organic Valley 2015).

Perishability has long been one of the main difficulties in marketing milk (DuPuis 2002), and ultra-pasteurization partially eliminates this obstacle to efficient, wide-scale marketing and sales. It seems that UHT processing has also contributed to firm concentration; there are fewer plants with UHT capacity, and it enhances the market power of national companies by facilitating the creation of geographically dispersed yet economically concentrated distribution chains. With fewer plants having UHT capacity, smaller companies with a more limited geographical reach are less likely to have access to one. With less concern about perishability as they ship product long distances to stores and warehouses, the national firms are well positioned to edge out regional firms. This interplay between market dynamics, technology, and distribution patterns leads to relatively few competitors in the organic milk distribution industry. The relative lack of competition contributes to an acceleration of the treadmill as this noncompetitive sector invests in more capital-intensive strategies to increase production and profits. However, this highly concentrated distribution node of the chain interacts in a very unbalanced manner, with more than 2,000 organic dairy farms that operate in a highly competitive environment, and whose survival largely depends on slowing down production per cow at the farm level.

Organic standards (e.g., prohibition against using antibiotics and requirement to only feed cows organically certified feed) have led to higher production costs and prices, which in turn have led to more dispersed distribution patterns in comparison to conventional milk. Unlike the conventional milk market, whose pricing system leads to clustering of producers near processing plants and population centers, the organic milk pricing system encourages a more dispersed system of production and distribution. While conventional milk market farm gate prices vary with distance from primary cities to reflect shipping charges to the plant and distance to market (Bailey 1997), organic milk farm gate milk prices do not vary according to proximity to market; they only vary by region to adjust for regional differences in production costs.

Firm Concentration and Branding in a Buyer-Driven Commodity Chain

Increasingly, conventional milk in supermarkets is sold under store brands, with regional milk brands, which often come from the exact same processing plant and thus the same milk pool, selling for 10 cents more a half gallon. With organic, the branding process has evolved such that the label "organic" with the "USDA Organic" seal carries enough cachet to justify the expense of national marketing campaigns. The brand recognition produced by these campaigns allows the national firms to charge higher prices than more regional or local organic dairy firms (King et al. 2010), further accelerating the dichotomous treadmill paths at the farm and marketing/distribution levels, respectively.

Within organic dairy a combination of legal, technology, and market factors have contributed to high levels of concentration at the firm level, with two firms buying and marketing 75 percent of the organic milk produced nationally. While it is certainly possible for more regional companies to succeed in this market, and there are more than fifty regional organic dairy firms, with companies such as Strauss Organic Creamery selling throughout the mountain West; and Springfield Creamery, based in Eugene, Oregon, selling its organic yogurt on the West and East Coasts, there appears to be a scale bias toward large firms with extended distribution and marketing reach. The small niche character of this industry has led to geographically dispersed distribution, which in turn supports market concentration. Because organic only has 5 percent of the US fluid milk market, it takes a population base twenty times greater than the conventional milk sector needs to absorb a given amount of production. Furthermore, the marketing driven nature of the industry supports concentration as larger firms have more resources to build brands and conduct expensive marketing campaigns.

It is necessary to dramatically extend the geographic extent of the distribution chain to capitalize on the branding process and build the company beyond the level of a small business. The standards partially limit farm size even as they contribute to concentration of economic activity at the firm or marketing level.

Organic Standards and Slowing the Treadmill on the Farm

While organic standards have no scale criteria, field research conducted by the author in the Northeast and Midwest indicates that there is an implicit scale bias in the standards, even before

FIGURE 5.8 *Organic, pastured Jersey cows in field. Photo courtesy of Janet Chrzan.*

the pasture standard was tightened up in 2010 to deal with complaints of an uneven playing field and uneven enforcement of the "access to pasture" provision of the National Organic Rule. If organic dairy farmers are required by their certifiers to pasture their animals it is very difficult to have an organic herd bigger than 500–600 cows. Farmers and key informants repeatedly articulated how difficult it is to pasture intensively large numbers of cows for three reasons: lack of contiguous pastureland, too much time spent taking cows to pasture and back, and too much energy spent by cows walking to and from the pasture. The pasture requirement seems to place a de facto upper limit on herd size among organic producers.

This scale bias of intensive pasturing slows down consolidation at the farm level and slows down the treadmill of production; pasture-based dairying systems require less machinery as field crop acreage is reduced or eliminated, and they have lower milk production per cow as animals spend more time ingesting a given level of caloric intake on a pasture-based diet compared to a grain diet, limiting their energy intake and their milk production, while reducing stress on their bodies as their metabolism slows down. One farmer explained this process by saying:

> I'm a grass farmer. I let the cows do the work. I haven't trimmed a cow's hoof in ten years. Cows are healthier and live longer on pasture.

One study of organic dairy farms in the Northeast found that for twenty-four farms reporting current and peak milk production (Diamond 2013), milk production per cow declined on average by 19 percent and grain feeding declined 43 percent from peak production to post-organic transition. While these figures overstate the impact of organic certification as the interim step of moving away from a confinement system toward a pasture-based system was a major contributor to reduced milk yields and less grain feeding, the shift to organic production reinforces this deceleration of the treadmill of production; the high price of organic grain makes heavy grain feeding uneconomical and risky because high per cow milk production increases the risk of mastitis, a common infection of the cow udder which is traditionally treated with antibiotics. Because organic standards completely prohibit antibiotic use on cows if their milk is ever to be marketed as organic, farmers aim to minimize mastitis by reducing average milk production per cow, which reduces stress on their cows' metabolic systems. The energy input for the cows is reduced, as is the product output; higher per unit prices are supposed to compensate for this reduction in bovine productivity.

Complementing this shift in bovine metabolism, dairy production under an organic regime appears to favor smaller farmers, and serves to value a more craft style of production over larger-scale, higher-input industrial production (Harper 2000). Average dairy size for both conventional and organic herds varies considerably by region, but regardless, organic herds are smaller than their conventional counterparts within the same region. Looking at the top four organic states: for California—organic herds average 507 cows compared to 1,214 for conventional herds; for Wisconsin—65 and 129; for New York—58 and 128; and for Oregon—382 and 521, respectively. Within dairying generally, consolidation of farms combined with higher production per cow are the two primary indicators of an accelerating treadmill of production, and with organic production the trend is clearly away from larger herds and higher input use/higher production per cow.

However, this slowing down of the treadmill of production on the farm is paralleled by and exists in tension with the acceleration of the treadmill of production downstream from the farm in the processing and distribution nodes of the commodity chain. Organic dairy farmers tend to replace purchased feed with on-farm sources, experiment with cow genetics to optimize grazing, and develop local sources of feed grain. On the distribution side large national milk firms dominate the market, and contribute to a cost price squeeze for farmers (Guptill 2009).

The creation of national brands of organic milk highlighting more "natural" production processes has contributed to the development of economically centralized and geographically dispersed distribution and marketing networks. These two factors in turn have put organic dairy farms in a vulnerable position. High marketing and distribution costs mean that less money is available for farm gate milk prices, while high organic grain prices raise production costs. This cost price squeeze has reached the point that one study indicated declining profitability for Vermont organic dairy farms from 2006 to 2010 (Parsons and McCrory 2011). In some cases, financial pressures have caused organic dairy producers to either exit farming or abandon organic certification (Maltby 2012). Nonetheless, organic certification has contributed to dramatic changes in production practices away from maximum production, high stress on bovine metabolic systems, and dependence on purchased inputs. How organic dairy firms respond to the cost price squeeze facing their farm suppliers will have significant implications for both the financial success of their suppliers, how they run their farms and the efficacy of organic certification as a tool for environmental change in the dairy sector.

Notes

1 While certainly organic advocates, farmers, and consumers would argue that food grown organically is healthier and better for the environment, the USDA has not taken this stance. The National Organic Program was placed within the Agricultural Marketing Service. When the national organic standards were first released in 2000, the Secretary of Agriculture at the time, Dan Glickman, stated, "Let me be clear about one thing, the organic label is a marketing tool. It is not a statement about food safety. Nor is 'organic' a value judgment about nutrition or quality" (www. webmd.com/food-recipes/news/20001220/new-usda-label-will-clearly-identify-organic-foods#1).

2 https://quickstats.nass.usda.gov/ is a searchable database through which one can make customized queries regarding practically any aspect of food and farming activity in the United States, going back decades, spanning the country or drilling down to counties, and encompassing all of agriculture, or specific subsectors.

3 The most complete data on the number of organic dairy farms in the United States comes from the 2014 Organic Production Survey (National Agricultural Statistics Service 2014) conducted by the USDA. For numbers of farms contracted with by CROPP and Horizon Organic, seewww.farmers. coop/pools/dairy-pool and www.horizonorganicfacts.com/facts-list.html

References

Bailey, K. W. 1997. *Marketing and Pricing of Milk and Dairy Products in the United States*. Ames, IA: Iowa State University Press.

Blayney, D. 2002. *The Changing Landscape of U.S. Milk Production*. Washington, DC: US Department of Agriculture. Bulletin number 978.

Clark, G. L., and Wrigley, N. 1997. The Spatial Configuration of the Firm and the Management of Sunk Costs. *Economic Geography* 73: 285–304.

Cochrane, W. 1993. *The Development of American Agriculture: A Historical Analysis*. Minneapolis, MN: University of Minnesota Press.

CROPP/Organic Valley. 2015. *Annual Report*. La Farge, WI: Author.

Diamond, A. 2013. Treadmill Acceleration and Deceleration: Conflicting Dynamics within the Organic Milk Commodity Chain. *Organization & Environment* 26: 298–317.

DuPuis, E. M. 2002. *Nature's Perfect Food: How Milk Became America's Drink*. New York: New York University Press.

Gereffi, G., Kornzeniewicz, M., and Kornzeniewicz, R. 1994. Introduction: Global Commodity Chains. In *Commodity Chains and Global Capitalism*, ed. G. Gereffi and M. Kornzeniewicz, 1–14. Westport, CT: Greenwood Press.

Goodman, D., and Redclift M. 1991. *Refashioning Nature: Food, Ecology, and Culture*. London: Routledge.

Gould, K., Pellow, D., and Schnaiberg, A. 2003. Interrogating the Treadmill of Production: Everything You Wanted to Know about the Treadmill, But Were Afraid to Ask. *Organization and Environment* 17: 296–316.

Gould, K., Pellow, D., and Schnaiberg, A. 2008. *The Treadmill of Production: Injustice and Unsustainability in the Global Economy*, Boulder, CO: Paradigm.

Gould, K. A., Schnaiberg, A., and Weinberg, A. S. 1996. *Local Environmental Struggles: Citizen Activism in the Treadmill of Production*. Cambridge: Cambridge University Press

Guptill, A. 2009. Exploring the Conventionalization of Organic Dairy: Trends and Counter-Trends in Upstate New York. *Agriculture and Human Values* 26: 29–42.

Guthman, J. 2009. Unveiling the Unveiling: Commodity Chains, Commodity Fetishism and the "Value" of Voluntary, Ethical Food Labels. In *Frontiers of Commodity Chain Research*, ed. J. Bair, 190–206. Palo Alto, CA: Stanford University Press.

Harper, D. 2000. Requiem for the Small Dairy: Agricultural Change in Northern New York. In *Dairy Industry Restructuring: Research in Rural Sociology and Development, Volume 8*, ed. H. K. Schwarzweller and A. P. Davidson, 13–45. New York: JAI Press.

Hill, H., and Lynchehaun, F. 2002. Case Study: Organic Milk: Attitudes and Consumption Patterns. *British Food Journal* 104: 526–42.

Hill, M. 2012. Organic Milk Low as Demand Up and Farmers Struggle. *Associated Press*, February 16. www.cnsnews.com/news/article/organic-milk-low-demand-and-farmers-struggle

Horizon Organic Dairy Company. 2001. *Annual Report*. Broomfield, CO: Author.

King, R. P., Hand, M.S., DiGiacomo, G., Clancy, K., Gomez, M. I., Hardesty, S. D., Lev, L., and McLaughlin, E. W. 2010. *Comparing the Structure, Size, and Performance of Local and Mainstream Food Supply Chains*. Washington, DC: US Department of Agriculture.

Lyson, T., and Geisler, C. C. 1992. Toward a Second Agricultural Divide: The Restructuring of American Agriculture. *Sociologia Ruralis* 32: 248–63.

Lyson, T., and Gillespie, G. W. 1995. Producing More Milk on Fewer Farms: Neoclassical and Neostructural Explanations of Changes in Dairy Farming. *Rural Sociology* 60: 493–504.

Maltby, E. 2006. Update on Pay Price. *NODPA News* 6: 1.

Maltby, E. 2012. Organic Dairy Family Farmers Need a Fair Share of the Retail Dollar, They Require At Least a 40 Cent a Gallon Increase in Their Milk Price to Stay in Business. *NODPA News* 12: 1.

Michaelis, K. 2013. Just Say No to UHT Milk. *Food Renegade*. www.foodrenegade.com/just-say-no-to-uht-milk/

Mooney, P. 1988. *My Own Boss? Class, Rationality, and the Family Farm*. Boulder, CO: Westview Press.

National Agricultural Statistics Service. 2014. *Census of Agriculture*. Washington, DC: US Department of Agriculture.

Novek, J. 2003. Intensive Hog Farming in Manitoba: Transnational Treadmills and Local Conflicts. *Canadian Review of Sociology and Anthropology* 40: 3–26.

Obach, B. K. 2007. Theoretical Interpretations of the Growth in Organic Agriculture: Agricultural Modernization or an Organic Treadmill? *Society and Natural Resources* 20: 229–44.

Parsons, B., and McCrory, L. 2011. Study Finds Declining Profitability of Vermont Organic Dairy Farms from 2006–2010. *NODPA News* 11: 1–7.

Richardson, R. 2008. Stonyfield & HP Hood Refuse to Pay Organic Dairy Farmers a Fair Price for Their Milk. www.organicconsumers.org/articles/article_10787.cfm

Schnaiberg, A. 1980. *The Environment: From Surplus to Scarcity*. New York: Oxford University Press.

Schnaiberg, A., and Gould, K. A. 1994. *Environment and Scarcity: The Enduring Conflict*. New York: St. Martin's Press.

Stevenson, S. 2009. Values-Based Food Supply Chains: Organic Valley. www.agofthemiddle.org/pubs/ovcasestudyfinalrev.pdf

Sturgeon, T. J. 2009. From Commodity Chains to Value Chains: Interdisciplinary Theory Building in an Age of Globalization. In *Frontiers of Commodity Chain Research*, ed. J. Bair, 110–135. Stanford, CA: Stanford University Press.

Su, Y., Brown, S. and Cook, M. 2013. *Stability in Organic Milk Farm Prices: A Comparative Study*. Washington, DC: Agricultural & Applied Economics Association Annual Meeting.

Swaminathan, N. 2008. Why Does Organic Milk Last So Much Longer Than Regular Milk? *Scientific American*. June 6.

Watts, D. C. H., Ilbery, B., and Maye, D. 2005. Making Reconnections in Agro-Food Geography: Alternative Systems of Food Provision. *Progress in Human Geography* 29: 22–40.

Weinberg, A. S., Pellow, D. N., and Schnaiberg, A. 2000. *Urban Recycling and the Search for Sustainable Community Development*. Princeton, NJ: Princeton University Press.

WhiteWave. 2014. *Annual Report, Form 10-K*. Denver, CO: Author.

6

Profile: Leigh Bush, Food Anthropologist

When I was in college I spent a few years working at a series of bakeries and coffee shops, thinking about whether or not I wanted to pursue a career in the kitchen or behind the counter. During the dead of winter, I would wake up at 4:00 a.m., driving with my head out the window, frozen air numbing my cheeks, because I didn't want to get up any earlier to scrape my windshield. When I arrived, I would fire up the ovens; retrieve the croissants, danishes, and cinnamon rolls that had proofed overnight; and bake off all the fresh pastries for that morning. Then, before anyone else arrived, I would make myself some coffee.

I had always liked coffee, though I wouldn't call myself a connoisseur. I came to discover, that just like everything we consume, coffee has a deep and long history going back to Ethiopia and spreading out globally over the course of a millennia. Through "coffee cuppings" I learned how to appreciate and taste the nuances of terroir, the regions where the coffee plant was grown, and the multiple methods by which the coffee cherry can be processed. Our roaster, who had visited all of the farms from which he sourced coffee, showed us the labor that goes into growing, processing, and roasting consistent and delicious beans. To achieve this quality, he gave his farmers an advance and guaranteed to purchase the beans at a fair price when they were grown. In exchange the farmers were able to grow coffee with the equipment, labor, and attention needed to produce high-quality beans.

After I applied to graduate school I traveled the world, working at and visiting farms, vineyards, and plantations. I was welcomed in by people who lived on boats made from plywood, in unfinished cinderblock and rebar buildings, and in dwellings dug into caves at the ocean's edge. Everywhere I went was different, but one thing remained common among the diverse groups of people I met: I was always served a beverage to welcome me into a space. I drank salty milks, aromatic chais, and mint teas so sweet they tasted like hot syrup. I swallowed gritty Turkish coffees and accepted instantly dissolving espresso powders. These beverages, I learned, were essential to human relationships—no matter how poor or wealthy, every culture managed to have some shared culture of drinking together.

When I arrived at graduate school the following year, I had the privilege of taking a class with a professor who works closely with coffee farmers (Catherine Tucker, who also has contributed

FIGURE 6.1 *Leigh in a commercial kitchen. Photo courtesy of Leigh Bush.*

to this volume with a chapter on the marketing of fair-trade and organic coffees). While I had met farmers and coffee drinkers across the globe, I had no idea that there are twenty-five million coffee farmers in the world, many of whom earn barely a dollar a day. Meanwhile, coffee is the second most valuable commodity traded on world markets. Coffee farmers, some of the poorest people in the world, also live in some of the most biodiverse and endangered habitats, where coffee production coexist with forests by retaining shade and native trees. This method of farming, however, is being overwhelmed by a drive for higher yields that favor sun-grown coffee, which requires forest clearing and toxic agro-chemicals. Furthermore, over the last several decades the countries that produce coffee went from retaining 20 percent to only retaining a mere 2 percent of the cost of a cup of coffee sold in a coffee shop. This is to say, as coffee prices rise, less money is given back to its producers.

Why do we treat something that so many people depend on, whether to feed their family or start their day so frivolously, I wondered? As much as I loved being a baker, and later, a barista, and appreciated the commitment my own roaster had to buying fair-trade coffee, I was confounded that I knew more about pouring a perfect microfoam rosette in a cappuccino than I did about

FIGURE 6.2 *Leigh with a stingray. Photo courtesy of Leigh Bush.*

the labor and systems that went into getting that bean into my grinder. As you read this chapter consider how the current system works to further create wealth disparity and entrench poverty. Ethiopia has several models of farms, all of which experience many of the same challenges faced by other organic farming systems. While conversion to organic practices has the potential to restore local ecology and increase yields, roadblocks like limited institutional support for useful technologies, access to markets, fair pricing, long-term affordable land leasing (at a size that can sustain a household), and advisories about best practices prohibit farmers from making these transitions. Without the support of consuming countries, coffee farmers have been unable to counteract the market domination of transnational corporations, resulting in an overproduction of low-quality coffee and a continued lack of equity and social justice for the more than twenty-five million coffee farmers. By getting to know your coffee roasters and buying fair-trade, consumers can demand that companies pay a better price to coffee workers, that they reduce waste and avoid chemicals, and that they work to locally address poverty and other social problems in coffee producing regions. How would you feel drinking that cup of Joe?

FIGURE 6.3 *Leigh in India. Photo courtesy of Leigh Bush.*

Going Organic: A Critical Analysis of the Potential for Organic Farming in Ethiopia

Bernhard Freyer and Jim Bingen

Introduction

Peasants in tropical and subtropical countries confront numerous interrelated internal and external challenges.[1] Drawing on our fieldwork in Ethiopia, we argue that organic farming offers several systemic solutions to address some of these challenges, but also needs further development. Without question, sustaining smallholder farming will require many fundamental transformations in the agrofood system as a whole and in their institutions and organizations. But, in this chapter we discuss how organic farming can be understood as a means to contribute to the sustainable development of smallholder farming in general.

Our study took place in three regions around Dangla (rain-fed) and Merawi (Koga irrigation scheme)/Amara Region and in the Awasha Region (semi-shadow coffee production) that illustrate many of the practices and challenges confronting smallholder farms in Ethiopia. From January to May 2016, we conducted a situation analysis (farm interviews, field walks, and field observations) on ten smallholder farms.

First we analyzed the household conditions that significantly influence the adoption/ transformation pathways of the smallholder farms. Based on our fieldwork, we constructed three smallholder model farms—local, high input, and organic—in terms of their materialities and techniques that serve as the decision situation for farmers adapting any practices now and in the future (Farr 1985). We sought to make visible the challenges smallholder farmers confront in establishing an environmentally friendly and economically viable farming system. After addressing several institutional surroundings of farming (e.g., religion, rural development, research, policies, etc.), we conclude with some observations about the complexity of a system change to organic farming.

FIGURE 6.4 *Farmer between tradition (hand hoe) and modernity (mobile telephone). Photo courtesy of Bernhard Freyer.*

Farm Household

Many family farms in Ethiopia are approximately 1.5 ha (3.7 acres) in size with a mixed farming system of crops and animals and some specializations, for example, honey, vegetables, fruits, or coffee. In general, households have shifted their materiality[2] from the use of on-farm resources and local origin products to the use of industrial products mostly from outside the country. This shift from traditional to modern material use (see Figure 6.4) in the household influences the organic matter cycle and energy production of the farm that in turn significantly influences the biodiversity and ecological status of the farm. It signals important changes in the productivity of the farm, for example, the closeness/dependency/self-sufficiency of the farm household in terms of materials and energy, and finally also their financial budget. We will show that this shift of materiality in the household involves changes in land use and the farm economy, as well as in labor and sociocultural patterns, and must be reflected when it comes to the question of how to convert a farm to organic.

Houses: architecture and furnishings. Families live in traditional wood and mud houses that are partly roofed with corrugated iron sheet, a mixture of traditional materials mainly from the farm (clay, construction wood, sometimes stones, bamboo, grass mats). Among the household furnishings, one finds some wooden chairs and usually some type of small table—all increasingly replaced by Chinese plastic furniture. Broadly speaking, the former use of locally produced household and kitchen utensils of wood and clay has been replaced—and with this, a decline in an income source for village artisans.

Water and sanitary systems. Drinking water is from wells or other groundwater sources that are at some distance from the house. Time-consuming activities significantly affect the availability of labor time that could be used to pursue possible innovations on the farm.

Energy systems. Firewood is harvested on the farm with a machete or, if limited on the farm, is purchased in the village or across the road. Also charcoal is used as well as dung as cooking fuel (Awasthi et al. 1996). In the meantime, the farming system is deficient in terms of farmyard manure. These energy approaches of the farm are hugely inefficient; the systems demand considerable work thereby limiting the labor capacity on the farm.

The social dimension of processing. All of the household practices are done by hand and are relatively labor-intensive. These practices help to create and maintain social relationships specifically among women. Any so-called modernization of this system with technology or materiality from outside the farm will affect these social interactions. In the words of practice theory, materiality affects the social.

Conclusion. The farm household described is on the edge between traditional to modern practices. But, even as the "plastic" and "sheet metal" culture is on the rise, the majority of materiality is still dominated by clay (mud), wood (see Figure 6.5) or cow dung for cooking—all the elements of a closed material cycle.

Farm Models

The studied farming systems are diverse in their agroecological, climate, soil, and socioeconomic conditions. But the commonalities allow us to classify them into three smallholder farm models and to assess the current sustainability status for each model.

Case Study #1: Local Farm Around Dangla

This farming system is found in many villages in the Amara region.[3] In general, it includes farms that are on the brink of using some modern agricultural techniques, despite being confronted with numerous limitations.

The farming system includes several interrelated subsystems: arable land; communal pasture land mainly outside the farm; the arable land using grain straw residues; diverse animal husbandry systems; some trees; and beehives (organic production). External inputs into these systems include small amounts of mineral fertilizer and feedstuffs (residues from local breweries or nug (*Guizotia abyssinica*) oil seed cake), as well as animals. Outputs are cereals, heifers, honey and some milk and meat, primarily for subsistence or for the market during the non-fasting season.

Crop production, pasture, and agroforestry

Farmland is used mainly for crop production (yields: maize 3.5 t/ha, teff 1.2 t/ha, wheat 2.5 t/ha), while agroforestry is limited to some trees.[4] These dense cereal-cropping systems produce low belowground root biomass. External inputs include mineral fertilizers (approx. 50 kg nitrogen/ha, DAP with 12 kg phosphorus/ha, herbicides and occasionally, pesticides mainly in maize, when they

FIGURE 6.5 *Gathering firewood. Photo courtesy of Bernhard Freyer.*

are affordable). Field practices include intense oxen ploughing (up to seven times), hand sowing, and hand hoe weeding. There is access to hybrid seed; however, proper handling of seeds is limited due to missing sowing technology. Herbicides may be used occasionally. Crop rotation does not include any humus enriching crops. Animals pasture all organic matter residues of arable land (leading to soil compaction and low humus content). There is little or no pasture management on communal land and thus, a loss of animal manure. As a result, pastures are overused and eroded. Farmyard manure, largely dried cow dung, is used primarily for cooking. Given the scarcity of cooking firewood, this practice is understandable. But it does contribute to increasing soil erosion (see Figure 6.6) and to the declining soil humus, low pH, and mineral depletion that all negatively affect soil quality and productivity. New practices that could address the problem, for example, the production of biogas, are found on a very limited number of farms. Also most farms have very few trees and shrubs.

Animal husbandry

Farms have three to five times more animals than the carrying capacity of the on-farm fodder production. Animals are undernourished, and the sale of perhaps one heifer a year contributes little economically to the household. As found in other studies, with an average of less than 1,000 kg per lactation,[5] milk production of indigenous breeds but first of all the lack of forage limits their growth potential (Mersha 2007). As also confirmed by Yilma et al. (2011: 26), natural grass hay, crop residues, teff and maize straw, stover of cereals and pulps, and agro-industrial by-products from the flour/oil industries and brewery residues are the main sources of fodder.

Honey production

To generate income, farmers have both traditional and modern beehives (4–6 kg and >40 kg honey per year per beehive). They collect and process the honey with a minimum of techniques. The honey is processed, packed by a Union[6] in Bahir Dar that sells the honey as organically certified, mostly for bakery use in Italy and France.

Post-harvest management, access to markets, and farm income

Cleaning and drying of cereals is done on a sheet of plastic in the sun. The grain is often rained on, thereby creating food losses and fusariosis[7] (Suleiman and Kurt 2015: Tsedaley and Adugna 2016). Cereals are stored in plastic bags that create the conditions for a high susceptibility to Fusarium mold and that do not prevent insect attacks or animal predators.

Most farmers are not directly linked to markets. Instead, brokers or a cooperative/union market products and create networks with customers. Thus, farmer information about market potential is limited and most must accept the offered price.[8] In other cases farmers travel significant distances to market their cereals, vegetables or animals. Even if they have access to a market, it may take days for them to sell their product, and they commonly lose money or even return home with all their products. Overall, there is limited human capital or political/financial power for change, a lack of knowledge and information to optimize the storage and market network; current material, economic, and social practices are kept in place.

Farm income does not allow investments in machinery or investment in higher education for the children. Financial reserves to cope with low harvests, illness, or the inability to work do not exist. On the other hand, these farmers produce enough food for their families.

Advisory services and research

Information about new or improved agricultural practices, or research activities, except in model farms, is limited. Extension services and relationships among neighbors tend to preserve long established practices. Even the advisory service repeat widely known recommendations for mineral fertilizer, herbicides and pesticides or the use of improved seed material. Human capital is limited and based on knowledge and experiences of the older generation and the neighbors (cultural capital). Most farmers either do not have access to credit or are unwilling to risk investing in inputs such as mineral fertilizer, especially at a 12–18 percent interest rate.

Conclusions

To sum up, as a result of crop production methods and a limited investment in agroforestry, the highly eroded soils have low levels of fertility with low humus content and low water holding capacity. In addition, insecure land tenure negatively influences the farming practices (Holden and Yohannes 2002) and hinders long-term investments such as liming or tree planting in soil fertility.

This farm type clearly does not depend upon external inputs. Nutrient cycles are open and nutrients are lost via ongoing water and wind erosion that negatively affects crop yields as well as both milk and meat production. The materialities of the hand hoe, the oxen plough (see Figure 6.9), or hinny (animal for transportation), conserve traditional practice formations.

Case Study #2: Intensified Farm in Merawi/Koga Irrigation Scheme

This farm model is privileged due to its location in an irrigation system. Since this type follows post-harvest management and animal husbandry activities, and has similar market access issues as the local farm type discussed above, we focus the discussion here mainly on crop production, agroforestry, farm income and use, and advisory services and research.

Crop production and agroforestry

Access to irrigation water intensifies crop production and allows two to three harvests per year.[9] Farmland is cropped with cereals, potato (maize, 4 t/ha, teff, 1.5 t/ha, wheat, 2.2 t/ha; potatoes, 8 t/ha; finger millet, 2.4 t/ha; barley, 2 t/ha), peas, beans and vegetable production. To increase income, farmers produce grain (+5 cents per kilogram) and onion seed material. Similar to the local farming system, cleaning and storage facilities that could protect cereals or potatoes against rain are lacking.

Fields are oxen-plowed and most of the field tasks—sowing, fertilizing, pesticide and herbicide spraying, harvesting and threshing—are done by hand. Organic manure or compost techniques are rarely used. The officially recommended amount of mineral fertilizer is about 100 kg nitrogen/ha (urea), DAP with 46 kg phosphorus and 17 kg nitrogen/ha per crop. But to realize two to three harvests per year, fertilizer and pesticide use must be higher than officially recommended. Onions are sprayed up to five times with fungicide. Crop rotation is not practiced. Most farmers are not trained to produce vegetables, nor do they have access to markets for large quantities of produce.

Furrow irrigation and heavy rains lead to serious fertilizer losses and contamination of groundwater water with herbicides, pesticides and nitrate. Low product quality and high pesticide residues in the soils and in the crops, specifically onions are critical issues.

More than in the case of the local farm model, the landscape includes few non-agricultural habitats. Hedges, agroforestry or alleys (see Figure 6.7) are an exception. Trees, if grown, are primarily eucalyptus woodlots.

Access to markets and farm income

Farm income allows farmers to own and use mobile telephones; the mobile increases their opportunities to receive information on fertilizers, pesticides, and the market situation. But, this is

FIGURE 6.6 *Erosion. Photo courtesy of Bernhard Freyer.*

no guarantee of environmentally and economically adequate applications of mineral fertilizers and pesticides.

Advisory services and research

The advisory service focuses on the irrigation scheme. But, the idea of a more environmentally friendly farming system based on crop rotation and organic matter management is neither offered by research nor by the regional advisory services. Furthermore, irrigation water is mismanaged and there is significant potential to increase the efficiency of water use.

Conclusions

While the irrigation scheme was originally reserved for vegetable production, a majority of the farmers are not vegetable growers. The dominating cereal production has led to important increases in yields and incomes. But it is coupled with the inefficient use of fertilizers and pesticides. In the long run these practices risk the creation of plant resistance and diseases. Due to the lack of soil organic matter, these practices will lead to soil degradation, compaction, acidification, and ground water contamination through agricultural inputs.

For smallholder farmers in the irrigation scheme with approximately 1.5 ha, this farming system has neither the financial nor the human capital for taking a technologically or an environmentally positive step forward. Going organic could be an option, but it would require a fundamental change in the production to include crop rotations that include forage legumes, cereals, grain legumes, potatoes, vegetables, and fruit trees, and mechanization, as well as organic manure and irrigation management.

Case Study #3: Mixed Organic Coffee Farm in the Awasha Region

Our third model is a mixed organic coffee farm in Awasha region where certified organic coffee is cultivated under half shade. It also includes several elements primarily of the first model (some crops, animals, fruit trees).[10] Here we again only highlight those subsystems of the farm that differ from those discussed above. Because this coffee farm model is an organic example, we specifically focus on the strengths and weaknesses of organic performances and techniques.

Crop production and agroforestry

Farm sizes range between 0.5 and 3.0 ha. Timber trees, some legume trees (e.g., *Sesbania sesban*) and Enset (false banana) provide half-shade. Some fruit trees (five to ten trees) like peach, apple, avocado, and papaya are cultivated in between, but the management is poor. The Enset leaf biomass serves for mulching and composting directly applied to the coffee plants.

The hand hoe is the main tool for managing the coffee, and ploughing is only done outside the coffee fields. In contrast to arable farming, the potential of intensification via mechanization is limited. The three cooperatives for coffee cleaning, pulping, and fermenting in this study did not systematically use the coffee husks for composting, thereby leading to a loss of organic matter and nutrients and a missed opportunity to close the nutrient cycle.

Post-harvest management, access to markets, and farm income

Consistent with national and international certification standards, the regional State Union collects the coffee, does the quality classification, and is responsible for marketing. Farm income through certified organic coffee is of relevance, but due to small farm size, the income and thus the potential for investment in the household and farm is limited.

Advisory services and research

Local as well as regional advisory services for coffee production provides some technical advice to farmers. But from an organic farming perspective there is little or no evidence of advice on a wide variety of significant agroecological issues: composting, the development of biogas, management to reduce post-harvest loss, training about nutrient cycles, liming and organic fertilizers, the intensification of the legume shrubs or cover plant management.

Conclusions

Coffee production demonstrates a way forward toward biodiversity, healthy soils, and plants. But natural resource management needs to be improved. Similar to the other farm types there is a lack of investment in technology and in several techniques that could improve productivity. We conclude that farm income based on approximately 2 ha in mixed coffee farms is enough to maintain the household. Even the higher organic prices for export coffee do not allow a serious step forward in their economic situation.

Lessons Learned

What are the commonalities and differences between the three farm models? We highlight the most relevant characteristics and outline some initial ideas for developing farming systems that are more sustainable and that could go organic.

Agroecology and crop production

Looking at the agroecological profiles of the farm models, there are obvious differences between the systems (see Table 6.1). The lack of organic matter management (e.g., cow dung, clover, alley farming) and low biodiversity is significant in the first and second models. In the slightly sloping region conditions of the (1–3 percent) first model farm, erosion is ongoing. At first glance, the irrigated farm of the second model presents a promising step toward modernizing the farming system. But the agroecological profile of this model is characterized by a loss of mineral fertilizers, the contamination of water with herbicides and pesticides, and low biodiversity. Soil fertility and organic matter management in the organic model is on the right track, and as coffee is a half-shade shrub, it is not surprising that by nature there is more investment into agroforestry than in the other models, but it requires further development with legume trees and under-sown legumes as well as proper coffee husk composting (Kassa et al. 2011).

Animal husbandry

Animal husbandry introduced in the first model, and also a feature in the second and third models has not changed since the analysis of Gryseels (1988) nearly thirty years ago: "shortage of feed during the dry season, lack of milk marketing facilities, particularly during the main fasting period, occasional disease problems of crossbred cattle, and the lack of appropriate breeding services."

TABLE 6.1 Agroecological performances of different smallholder farm models in Ethiopia

Characteristics	Local Farm	Intensified Farm*	Organic Farm
Soil fertility status	+	+	+++
Crop rotation/ intercropping	–	–	++
Erosion management	–	–	++
Agroforestry/alley farming	+	–	++
Hedges	+	–	++
Fruit trees	+	–	++
Mulching strategies	–	–	+++
Pesticide use	+	+++	–
Herbicide use	+	+++	–
Nitrogen fertilizer	+	+++	–
Phosphorous fertilizer	+	++	+
Liming	–	–	–
Farm yard manure/ composting	–	–	++ (1) / + (2)
Biogas	–	–	+

Key: – = negative; + = little, low; ++ = medium; +++ = high; *irrigation.
(1) farm; (2) coffee husks at the cooperative processing site.

Economy and markets

From an economic perspective, the local farm model with one organic product (honey), and the organic farm model with certified organic coffee with their linkage to the international market generate sufficient income to maintain the current living standard and food security. The second intense farm with two to three harvests under irrigation per year and high fertilizer input allows as well a basic income for the family.

What is the important economic difference between the organic and the nonorganic farms? The nonorganic purchase fertilizers, pesticides, and feed stuff inputs. But the organic rely on more human capital and additional labor to help provide the additional organic matter, manage the natural resources and improve the biodiversity (alleys, crop rotations).

Entry Points to Go Forward

For all farm types, the lack of financial and human capital hinders optimizing farming systems, for example, through investment in new labor saving techniques (Marenya and Barrett 2007). Financial capital does not exist for technical innovations that could fundamentally reduce the workload. Farmers continue their traditional practices with low inputs, and inadequate organic matter management that leads to environmentally negative impacts. System oriented innovations to increase biomass, like clover production, composting, or agroforestry and alley farming that would help to stop erosion and would contribute to a much higher milk and meat production with fewer animals, are unknown, ignored or applied sporadically and ineffectually.

What are their opportunities for breaking out of this cycle? The identified weaknesses are well known, and there is little or no move toward the use of what we describe as "good farming practices," for example, composting or adapted application of mineral fertilizers or crop rotations. Social and cultural capital, as well as the absence of financial capital, work to conserve current

FIGURE 6.7 *Tree alleys in farmland. Photo courtesy of Bernhard Freyer.*

practices and preclude fundamental changes in farming practices (see Bourdieu 2005: 63). In addition, many of the challenges to innovation lie outside the farm/household system, for example, the absence of advisory services and financial credit or supportive market regulations.

Discussion

Based on the above farm models we explore the opportunity and investment required for each farm model to convert to organic. This also sheds light on how conversion can be a means of adaptation at the farm level and the extent to which conversion requires changes in the institutional surroundings.

Farm internal potential and related challenges to go organic

Cropping and fertilizer systems

Challenges are the nitrogen and phosphorus shortfalls at critical moments in the growing season that limit growth rates and yields (Breman et al. 2001). In the long rainy season the critical issue is not the quantity of water, but efficient nitrogen and phosphorous application (Ruben and Heerink 1995). As Birech et al. (2008) note, a specific water supply is required to achieve an important yield, and to assure nitrogen fixation with legumes during the short rainy season. Today, we are able to refer to best organic matter practices to solve many of the problems in an ecological manner.
There is evidence that intercropping (Nnadi and Haque 1986; Fujita and Budu 1994; Mpairwe et al. 2002; Akande et al. 2006; Nedunchezhiyan et al. 2011; Dwivedi et al. 2015), the use of cow dung (Ayoola and Makinde 2008), and the use of forage legumes, for example, lablab, can compete with the use of mineral fertilizer (Birech et al. 2014).

Animal husbandry and feeding strategies

Reconfiguring the use of crop and pasture land will be central in each model. Overgrazing has led to significant yield declines in forage crops that can be compensated for only by cultivating improved forages (such as alfalfa, clover, Napier grass, etc.). But currently, these plants only cover 0.25 percent of the nutritional need of animals in Ethiopia (CSA 2010b) and 0.18 percent of animal feed needs in the Amhara Region (Firaw and Getnet 2010). Increased dairy productivity is linked to access to protein and starch from rich green fodder and hay from leguminous plants with Napier grass as a starch rich plant (CSA 2010a).
Animal traction and threshing are one of the main reasons for keeping cattle on each of the farm models. However, reduction of the number of animals is needed to avoid soil erosion and compaction both on arable and pasture land.

Labor and mechanization—compost and weed control as examples

Without exception, the farms have to cope with significant demands on household labor. Consequently, the additional labor required in the move to organic always raises critical questions. Alley farming and composting also demand a significant amount of labor.

Compost management and sprayers to reduce labor are only affordable with external financial support and with a cooperative approach through which farmers share the investment and maintenance costs of modern technology.

Investment into zero grazing combined with a pasture system would lead to an increase of farmyard manure. Compost sprayers and improved techniques for cutting and transporting clover from the crop rotation could be an investment by farmer groups or at a communal level.

Mechanized weeding with a horse-, ox-, or tractor-drawn weeder would significantly reduce the farm workload. Crop diversification with fodder legumes (mandatory for organic farming) can also reduce weed pressure.

Implementation phase of organic techniques

There is no question that these organic management methods need time to be successfully implemented on smallholder farms. Increased crop yields can be expected only after the second year from the following practices: direct pre-crop effects of legumes, the application of farmyard manure, and the use of cuttings from alley trees (see Figure 6.8). That is, this time gap between the investment in organic practices and the economic return during the conversion period presents a key challenge for a systems change. The delayed impact on income is one of the main hurdles keeping farmers from investing in organic farming, that is, intense organic matter management.

FIGURE 6.8 *Introducing farmers to alley farming. Photo courtesy of Bernhard Freyer.*

Several types of incentives would be needed to motivate farmers to move to organic, for example, high support through advisory services, and technical support.

Farm external conditions for organic farming development

Since several current institutional conditions negatively affect the development of agriculture as a whole, some recommendations must be oriented toward strengthening the agricultural sector in general, and not limited specifically to what is required to support organic.

Production

Farm size and land tenure. All three farming systems are confronted with limited land. An average between 0.5 and 3.0 ha (1.2–7.4 acres; 7.4 acres are only realized by a minority of farmers) is insufficient to generate an attractive income. Thus, even after converting to organic and further differentiation and optimization of the farm, there is need to increase farmland per farmer family combined with mechanization or off-farm income.

Given the uncertain land tenure conditions, many farmers are hesitant to invest in longer term practices to improve soil fertility, or in tree planting (IFPRI 2005: 75). Thus, going organic is a specific challenge since it requires long-term investments in soil fertility.

The potential for increasing farm size in the highly populated areas is limited. The differentiation and optimization of production is one option. Job diversification into processing and marketing of agricultural products is an option only for some. Other employment in the service sector or other industries remains limited.

Farm inputs and supply industries. While there are governmental programs that push the application of mineral fertilizers and pesticides, there are no programs to address basic soil fertility. Currently there is no supply of, or research on the use of clover, alfalfa, or desmodium seeds. Obviously, research and market development for organic inputs must be established.

Economy

Markets and certification. Abusive market behavior in the food distribution chain specifically challenges sustainable income generation for smallholder farmers (IFPRI 2005). Markets operate without governmental oversight or control of the brokers that could create the conditions for the development of fair market prices. IFPRI (2005) noted that farmers are not motivated to increase production if storage, transportation, and fair market conditions cannot be improved. Weatherspoon and Reardon (2003) observed that traditional retailers and farmers would further lose influence on the market if supermarket chains take over the control and define the rules including quality standards, which is evident in the North. If the state strengthens liberalization of the market without regulations for a minimum price, this would only lead to a society of some richer people while the majority stays poor. Consequently a market framework must be developed that fits with the production conditions, the farm economy as well as consumer financial capacities. So the question is: can market conditions that encourage smallholder production for the market be designed and enforced?

Participatory guarantee systems (Källander, 2008; AdeOluwa 2010) and similar approaches such as Fair Trade offer options for improving fair market conditions that could promote organic farming. The strict European food standards (e.g., technical quality of vegetable and fruit products) cannot be fulfilled by many smallholder farmers (see also Ruben et al. 2003), for example, requirements in terms of application of industrial organic fertilizers that are costly or organic matter management based on the farmers' own resources. Also, different understandings of the technological quality of products exclude a high percentage of farm products, which was the case with pineapple in Uganda (Jumba and Freyer 2016). In all cases there is need for intense research and training to support farmers in their conversion process toward real sustainable agricultural systems.

Adapted credit and subsidy systems. With credit at 15 percent interest, most farmers are precluded from investing in the farm and household, except some small tools that might help to facilitate the production but not really change the workload. Cooperative approaches allow alternative economic strategies.

Sociocultural issues

Migration. Current off-farm migration includes temporary migration to gain off-farm income, migration to urban centers, and migration to another country. Organic agriculture offers an

FIGURE 6.9 *Traditional ploughing. Photo courtesy of Bernhard Freyer.*

alternative because of the diversified labour demands for production, processing and marketing of agricultural products. Organic agriculture can also create additional employment opportunities through tourism as well as several ecosystem services that could attract young people and thereby help to reduce out-migration.

Religion. The Christian orthodox religion dominates in rural areas. Observing religious celebrations require more than 170 days per year. During these times farmers reduce or cease their fieldwork. This often conflicts directly with weather-dependent agronomic activities and jeopardizes timely sowing and harvesting. Under these circumstances, an investment in technology might partly allow for more timely fieldwork that does not compete with religious days.

Research and Extension

The advisory services and research centers focus on external inputs that only can—from their perspective—improve both land and labor productivity (Ruben 2005). They recommend the use of inorganic fertilizers, synthetic pesticides, and hybrid crop varieties. But there is limited evidence that these recommendations provide sustainable yields in the long term or avoid erosion and the decline of soil fertility and other environmental damages. Education and training will be the key to any step forward in the use of agroecological, that is, organic matter practices, more efficient production, technology implementation, and market development.

Conclusions and Recommendations

We have shown that several organic system technologies and practices can help to solve some environmental challenges and increase meat, milk, and crop production over the long term. Going organic means recognizing that this approach offers a means to address decades of natural environment damage. The analysis further makes clear that there is need for an intensification of the production. This will be not possible without investments in technology (e.g., adapted soil tillage, mechanical weed control, or composting techniques). Effective investment strategies that would make these technologies affordable for farmer groups are also needed.

From a theoretical point of view, we conclude that the move from traditional materials and techniques to modern resources ("material turn") and practices involves two pathways that can be differentiated in terms of the origin of materiality, the material type, and the related material cycle. The first pathway can be characterized as an input-based high-energy-consuming approach that we call "industrial modernization (IM)," where main materials and technologies are mineral fertilizers, herbicides and pesticides (industrial inputs). The materiality comes mainly from off the farm, is costly, has a significant ecological footprint, and is unrelated to the production scheme of the farm. Moreover, it does not build on local knowledge and experiences.

The second pathway is what we call the "reflexive ecological modernization (REM)" pathway.[11] It is based on the farm's organic materiality or its capacity to produce some of its materialities within its own system. As a precondition, it involves materiality given by nature, that is, high biodiversity, natural based pesticides and organic matter production on the farm. It includes organic farming practices combined with agroforestry and timber production, and leads directly to a closed energy system on the farm and a high carbon sequestration, using biogas and solar technology. To run the organic approach, investment into technology (local and industrial based) is a must.

While the first pathway tends to orient social interactions toward the input market and the local advisory and research services, the reflexive pathway integrates more local and regional sites of shared labour and social interactions due to the production and processing of organic matter.

This short description indicates that the material base of farmers has to be changed fundamentally, in order to stop the current devastation of land and to overcome low productivity. Ethiopian agricultural policy currently does not include organic agriculture as a strategy (Anonymous 2015). As mentioned above the mainstream approach can be characterized as a kind of second green revolution. The organic pathway also partly contradicts the currently dominant neoliberal policies and market strategies that are highly interested in agricultural investors with an export orientation. Therefore, an "organic" program is equal to a paradigm shift for Ethiopian agricultural policies, research, and extension and advisory services, as well as for countries with similar cultural, material, socioeconomic, and agroecological conditions.

FIGURE 6.10 *Traditional plough. Photo courtesy of Bernhard Freyer.*

NOTES

1 Farm internal challenges—*agroecology and productivity*: high soil erosion (Haileslassie et al. 2005; Tilahun et al. 2014), low yields (low farm management performance; Tittonell et al. 2005; Zingore et al. 2007), relatively simple production technologies (Ruben 2005), and lack of knowledge about sustainable agricultural practices and low technological standards; *economy*: limited land, limited financial resources, limited access to markets, lack of (non-farm) off-farm income (Barrett et al. 2001; Holden et al. 2004).

Farm external challenges—insecure land tenure conditions, high population development and poverty of farmers (Shiferaw and Holden 1998), HIV-AIDS (Drimie 2002) and other health problems (Russell 2004), hunger and temporary food shortage, and, most recently, climate change.

2 Materiality describes all kinds of materials that can be part of nature (plants, soils, animals, leaves) or man-made (hand hoe, fertilizer, pesticides, house, cooking stove, car), which together allow a certain practice, for example, to reduce weeds with herbicides or with hand hoe. If a certain materiality is (not) available, then a certain practice is (not) possible. Transformation of a farm toward sustainability depends on the availability of certain materiality, for example, alfalfa seeds that can fix nitrogen.

3 Climate data: The annual rainfall is 1.550 mm (http://en.climate-data.org/region/1502/?page=4) and more annually with one rainy season and temperature is around 16.8°C (Hurni 1998). In this region there is no more forest except some eucalyptus plantations. The landscape shows erosion gullies, and in every rainy season, erosive processes seriously affect the slightly inclined landscape.

4 Intensified organic or conventional production in the South allows maize yields between 6 and 10 t/ha. Under industrialized farming conditions in the North, more than 12t/ha maize can be produced.

5 Under optimized conditions in the South, 3–6 tons per cow is possible; in the North under industrial conditions, one cow can produce more than 10 tons of milk.

6 www.zembaba.bee.org.

7 *Fusarium spp.* is a widespread plant fungi surviving in soils and growing under monocropping and humid conditions all over the world. It infects all kind of plants (wheat, maize, vegetables, fruits) and produces a mycotoxin that is poisonous to humans and animals.

8 Honey is collected by a cooperative and processed by a union, responsible for the marketing and certification as an organic with a price of approx. 4 Euro/kg.

9 Climate data: The rainfall is 1,487 mm annually (Bewket 2009) with one rainy season and temperature is around 17.8°C. Also in this region there is no more forest except some eucalyptus plantations.

10 Climate data: The rainfall is approx. 965 mm with one rainy season and temperature is around 17.2°C.

11 Reflexive modernization represents an alternative to the industrialization of the environment. It considers local and global consequences of practices and the attitude "that we do not know" as an entry point to a structural learning process (Gill 1999). Reflexive ecological modernization is defined through an agriculture that tends to exclude environmental and health-risky farming methods and farm inputs.

References

AdeOluwa, O. O. 2010. *Organic Agriculture and Fair Trade in West Africa*. Rome, Italy: FAO.

Akande, M., Oluwatoyinbo, F., Kayode, C., and Olowokere, F. 2006. Response of Maize (*Zea mays*) and Okra (*Abelmoschus esculentus*) Intercrop Relayed with Cowpea (*Vigna unguiculata*) to Different Levels of Cow Dung Amended Phosphate Rock. *World Journal of Agricultural Sciences* 2: 119–22.

Anonymous. 2015. *The Second Growth and Transformation Plan (GTP II) (2015/16–2019/20) (Draft)*. Addis Ababa: Addis Ababa National Planning Commission; The Federal Democratic Republic of Ethiopia.

Awasthi, S., Glick, H. A., and Fletcher, R. H. 1996. Effect of Cooking Fuels on Respiratory Diseases in Preschool Children in Lucknow, India. *American Journal of Tropical Medicine and Hygiene* 55: 48–51.

Ayoola, O., and Makinde, E. 2008. Performance of Green Maize and Soil Nutrient Changes with Fortified Cow Dung. *African Journal of Plant Science* 2: 19–22.

Barrett, C. B., Reardon, T., and Webb, P. 2001. Nonfarm Income Diversification and Household Livelihood Strategies in Rural Africa: Concepts, Dynamics, and Policy Implications. *Food Policy* 26: 315–31.

Bewket, W. 2009. Rainfall Variability and Crop Production in Ethiopia: Case Study in the Amhara Region. Paper presented at the Proceedings of the 16th International Conference of Ethiopian Studies.

Birech, R., Freyer, B., Friedel, J., and Leonhartsberger, P. 2008. Effect of Weather on Organic Cropping Systems in Kenya. Paper presented at the 16th IFOAM Organic World Congress, Modena, Italy.

Birech, J., Freyer, B., Friedel, J., Leonhartsberger, P., Macharia, J., and Obiero, C. 2014. Biomass Production in the Short Rains and Its Influence on Crops in the Long Rains: A Systems Approach in Organic Farming. *American Journal of Experimental Agriculture* 4: 1047.

Bourdieu, P. 2005. Ökonomisches Kapital – Kulturelles Kapital – Soziales Kapital (Jürgen Bolder, Ulrike Nordmann & et al., Trans.). In *Die Verborgenen Mechanismen der Macht. Schriften zu Politik & Kultur 1*, ed. Margareta Steinrücke, 49–79. Hamburg: VSA-Verlag.

Breman, H., Groot, J. R., and van Keulen, H. 2001. Resource Limitations in Sahelian Agriculture. *Global Environmental Change* 11: 59–68.

CSA. 2010a. *Agricultural Sample Survey. Volume II*. Addis Ababa, Ethiopia: CSA Central Statistical Authority.

CSA. 2010b. *Statistical Abstract 2010*. Addis Ababa, Ethiopia: CSA Central Statistical Authority.

Drimie, S. 2002. *The Impact of HIV/AIDS on Rural Households and Land Issues in Southern and Eastern Africa*. A background paper prepared for the Food and Agricultural Organization, Sub-Regional Office for Southern and Eastern Africa. Pretoria, South Africa: Human Sciences Research Council.

Dwivedi, A., Dev, I., Kumar, V., Yadav, R. S., Yadav, M., Gupta, D., . . . Tomar, S. 2015. Potential Role of Maize-Legume Intercropping Systems to Improve Soil Fertility Status under Smallholder Farming Systems for Sustainable Agriculture in India. *International Journal of Life Sciences Biotechnology and Pharma Research* 4: 145.

Farr, J. 1985. Situational Analysis: Explanation in Political Science. *Journal of Politics* 47: 1085–107.

Firaw, T., and Getnet, A. 2010. *Feed Resource Assessment in Amhara National Regional State. Ethiopia Sanitary and Phytosanitary Standards and Livestock and Meat Marketing Program (SPS-LMM)*. Ethiopia: USAID.

Fujita, K., and Budu, K. 1994. Significance of Legumes in Intercropping Systems. Roots and Nitrogen in Cropping Systems of the Semi-arid Tropics . *JIRCAS International Agriculture Series* 3: 19–40.

Gill, B. 1999. Reflexive Modernisierung und technisch-industriell erzeugte Umweltprobleme-Ein Rekonstruktionsversuch in präzisierender Absicht. *Zeitschrift für Soziologie* 28: 182–96.

Gryseels, G. 1988. *Role of Livestock on Mixed Smallholder Farms in the Ethiopian Highlands: A Case Study from the Baso and Worena Wereda near Debre Berhan*. Landbouwuniversiteit te Wageningen.

Haileslassie, A., Priess, J., Veldkamp, E., Teketay,D., and Lesschen, J. P. 2005. Assessment of Soil Nutrient Depletion and Its Spatial Variability on Smallholders' Mixed Farming Systems in Ethiopia Using Partial versus Full Nutrient Balances. *Agriculture, Ecosystems & Environment* 108: 1–16.

Holden, S., Shiferaw, B., and Pender, J. 2004. Non-farm Income, Household Welfare, and Sustainable Land Management in a Less-Favoured Area in the Ethiopian Highlands. *Food Policy* 29: 369–92.

Holden, S., and Yohannes, H. 2002. Land Redistribution, Tenure Insecurity, and Intensity of Production: A Study of Farm Households in Southern Ethiopia. *Land Economics 78*: 573–90.

Hurni, H. 1998. *Agroecological Belts of Ethiopia*. Bern: Centre for Development and Environment, University of Bern in association with the Ministry of Agriculture, Ethiopia.

IFPRI. 2005. *The Future of Small Farms. Proceedings of a Research Workshop*. Wye, UK: International Food Policy Research Institute.

Jumba, F. R., and Freyer, B. 2016. Perception of Quality in Certified Organic Pineapples by Farmers in Kayunga District, Central Uganda: Implications for Food Security. *Journal of Agriculture and Rural Development in the Tropics and Subtropics* 117: 137–48.

Källander, I. 2008. Participatory Guarantee Systems—PGS. Swedish Society for Nature Conservation. www.ifoam.org/sites/default/files/page/files/pgsstudybyssnc_2008.pdf

Kassa, H., Suliman, H., and Workayew, T. 2011. Evaluation of Composting Process and Quality of Compost from Coffee By-products (Coffee Husk and Pulp). *Ethiopian Journal of Environmental Studies and Management 4*.

Marenya, P. P., and Barrett, C. B. 2007. Household-Level Determinants of Adoption of Improved Natural Resources Management Practices among Smallholder Farmers in Western Kenya. *Food Policy* 32: 515–36.

Mersha, S. Y. 2007. *Dairy Production, Processing and Marketing Systems of Shashemene—Dilla Area, South Ethiopia*. MSc thesis, Awassa College of Agriculture, School of Graduate Studies, Hawassa University, Awassa, Ethiopia.

Mpairwe, D., Sabiiti, E., Ummuna, N., Tegegne, A., and Osuji, P. 2002. Effect of Intercropping Cereal crops with Forage Legumes and Source of Nutrients on Cereal Grain Yield and Fodder Dry Matter Yields. *African Crop Science Journal* 10: 81–97.

Nedunchezhiyan, M., Laxminarayana, K., Rajasekhara Rao, K., and Satapathy, B. 2011. Sweet Potato (Ipomoea batatas L.)-Based Strip Intercropping: I. Interspecific Interactions and Yield Advantage. *Acta Agronomica Hungarica 59*: 137–47.

Nnadi, L., and Haque, I. 1986. Forage Legume-Cereal Systems: Improvement of Soil Fertility and Agricultural Production with Special Reference to Sub-Saharan Africa. In *Potential of Forage Legume in Farming Systems of Sub-Sahara Africa, ILCA, Addis Ababa, Ethiopia*, ed. I. Haque, S. Jutzi and P. J. H. Neate, 330–362. Addis Ababa, Ethiopia: International Livestock Centre For Africa.

Ruben, R. 2005. Smallholder Farming in Less-Favored Areas: Options for Pro-Poor and Sustainable Livelihoods. In *The Future of Small Farms: Proceedings of a Research Workshop*, 83–102. Washington, DC: International Food Policy Research Institute.

Ruben, R., and Heerink, H. 1995. Economic Evaluation of LEISA Farming. *ILEIA Newsletter* 11: 18–20.

Ruben, R., Kuyvenhoven, A., and Hazell, P. 2003. Institutions, Technologies and Policies for Poverty Alleviation and Sustainable Resource Use. Presented at Staying Poor: Chronic Poverty and Development Policy, Institute for Development Policy and Management, University of Manchester, 7–9 April.

Russell, S. 2004. The Economic Burden of Illness for Households in Developing Countries: A Review of Studies Focusing on Malaria, Tuberculosis, and Human Immunodeficiency Virus/ Acquired Immunodeficiency Syndrome. *American Journal of Tropical Medicine and Hygiene* 71 (2 suppl): 147–55.

Shiferaw, B., and Holden, S. T. 1998. Resource Degradation and Adoption of Land Conservation Technologies in the Ethiopian Highlands: A Case Study in Andit Tid, North Shewa. *Agricultural Economics* 18: 233–47.

Suleiman, R. A., and Kurt, R. A. 2015. Current Maize Production, Postharvest Losses and the Risk of Mycotoxins Contamination in Tanzania. Paper presented at the 2015 ASABE Annual International Meeting.

Tilahun, S. A., Guzman, C. D., Zegeye, A. D., Ayana, E. K., Collick, A. S., Yitaferu, B., and Steenhuis, T. S. 2014. Spatial and Temporal Patterns of Soil Erosion in the Semi-Humid Ethiopian Highlands: A

Case Study of Debre Mawi Watershed. In *Nile River Basin*, ed. A. Melesse, W. Abtew, and S. Setegn, 149–63. Cham: Springer.

Tittonell, P., Vanlauwe, B., Leffelaar, P., Rowe, E. C., and Giller, K. E. 2005. Exploring Diversity in Soil Fertility Management of Smallholder Farms in Western Kenya: I. Heterogeneity at Region and Farm Scale. *Agriculture, Ecosystems & Environment* 110: 149–65.

Tsedaley, B., and Adugna, G. 2016. Detection of Fungi Infecting Maize (Zea mays L.) Seeds in Different Storages around Jimma, Southwestern Ethiopia. *Journal of Plant Pathology and Microbiology* 2016.

Weatherspoon, D. D., and Reardon, T. 2003. The Rise of Supermarkets in Africa: Implications for Agrifood Systems and the Rural Poor. *Development Policy Review* 21: 333–55.

Yilma, Z., Guernebleich, E., Sebsibe, A., and Fombad, R. 2011. *A Review of the Ethiopian Dairy Sector*. Addis Ababa, Ethiopia: FAO Sub Regional Office for Eastern Africa.

Zingore, S., Murwira, H. K., Delve, R. J., and Giller, K. E. 2007. Influence of Nutrient Management Strategies on Variability of Soil Fertility, Crop Yields and Nutrient Balances on Smallholder Farms in Zimbabwe. *Agriculture, Ecosystems & Environment* 119: 112–26.

7

Profile: Erika Tapp, Concerned Consumer

My concern with what I'm eating started when I learned to cook. Like many others my age, I grew up in a house with a lot of snacks and processed foods; my mom didn't cook so much as assemble. I now really enjoy cooking and appreciate the better flavors of the local foods from the community-supported agriculture (CSA) and the farmers' market and I don't cook much that's processed. Getting to that point took a lot of steps and a lot of learning, from reading labels to thinking about where the food is grown, to wondering how it's grown.

A native Californian, I have been living in Philadelphia for sixteen years. It's been an exciting time to be here, since the restaurant scene has exploded and the farm to table movement has led the way. Philadelphia is so into its food, from the new farmers' markets that are popping up every year, to the local food sold in markets and even corner stores and new restaurants that proudly use local Pennsylvania and New Jersey foods. Everywhere you look there are community gardens and food plants in front yards. Many neighborhoods are becoming more economically vibrant and ethnically diverse, and leading the way are restaurants opened by new, experimental chefs who are taking advantage of cheap real estate to start food businesses. We are lucky to be located right in the middle of excellent agricultural land as well—New Jersey, Lancaster, Berks and Bucks counties, and even Maryland, which provides farm produce and seafood. So it's easy to be food and health conscious at this time and place, and I feel lucky to be able to source my food from so many local farmers and food makers.

I am an amateur foodie, and enjoying cooking. With my own cooking, I rarely use processed food (with the exception of things like pasta), though I certainly eat those in other situations. Vegetable and herbs are a large portion of what I plant. However, I readily acknowledge that it is more a hobby than a solution to feeding myself or my family.

In this chapter, Klimek et al. discusses the origins and values of the organic food movement, which embodied a holistic understanding of the ecological implications of organic farming. This history resonated for me. I see the disparity between the organic section at Walmart and the local farmers' markets. As a consumer, I want to support a sustainable and fair food system. Yet, I also get many of my groceries at the supermarket. I do my best to buy local and organic because I've always considered myself an environmentalist, though not adamantly so. I diligently recycle, compost when I can, pay extra for renewable energy for my home, take canvas bags to the

FIGURE 7.1 *Erika Tapp at the farmers' market. Photo courtesy of Janet Chrzan.*

grocery store, drive an efficient car, and donate to local organizations addressing the preservation of the environment or food insecurity. I am very concerned with environmental impact of farming and how our current farming practices and use of pesticides affect groundwater and soil quality. I'm also pretty worried about my carbon footprint and so I often prefer local over long-distance organic for environmental impact and nutritional value. I feel that our world is so full of dangerous and carcinogenic chemicals that what I'm getting from a tomato is the least of my concerns. Chemicals in beauty and cleaning products, air pollution, whatever is in my water, drinking alcohol, or simply not getting enough fruits and veggies are far more likely to affect my health long-term.

In addition, I hear varying reports from food and agriculture professionals that "organic" is a loose term and does not necessarily reflect true environmental sustainability or quality of product. Klimek et al. also explores the philosophical underpinnings of the organic movement, which extends beyond the specific issues of pesticides to include sustainability, workers' rights, and health. Those are the values that I connect with, and rely on when I am making my consumer choices.

For the last ten years I have worked in community-based organizations in North Philadelphia dedicated to community and economic development as well as social services. Much of my work has focused on quality of life for residents in underserved neighborhoods. Programming has included advocacy for responsible development, direct services to low-income residents and seniors, job training and placement, adult literacy, transitioning out of homelessness, small business support, food access, and community greening. Community greening is a broad term that encompasses developing community gardens and urban farms, tree planting and maintenance, promoting recycling and e-cycling.

I've seen how food leads toward community cohesion and gentrification in my work as well. I've seen firsthand how gentrification affects neighborhood residents, for good and for bad. The urban farms that hipster college grads want to start in poor neighborhoods too often can start a process that displaces long-term residents whose incomes are not enough to buy the organic food produced in those farms—or to keep up with increased real estate tax assessments. As a resident of Philadelphia, I see farmers' markets erupting like mushrooms in communities all over the city. There are concerted efforts to establish them in neighborhoods where there is a lot of food insecurity. The city and state have created various food subsidy programs providing extra dollars or incentives to shop at farmers' markets. Vendors are making sincere efforts to accept food stamps. However, a $5 tomato is still a $5 tomato. And when you only have $40 per week to spend on food, you're not spending $5 on a tomato, even if you get one dollar back. Similarly, in grocery stores, organic is more expensive for the consumer, full stop. Buying decisions for the food insecure are primarily driven by economics, secondarily by convenience.

FIGURE 7.2 *Erika Tapp cooking dinner at home. Photo courtesy of Janet Chrzan.*

In the neighborhoods I have worked in, food access is less and less of an issue. There have been many efforts to make sure that everyone has access to a grocery store. If our actual concern is for the health and wellness of our communities, then we need to focus more on equitable access as well as educating working families in how to use fresh ingredients, value "real" food over processed, and embrace the social and health benefits of preparing a family meal. But most importantly, parents need to be able to earn a living wage that allows them to be home in the evenings (as opposed to working second or third jobs), and to have jobs with family friendly policies. When local, sustainable, and healthy isn't economically accessible to all city residents it remains just another reminder of inequality.

Organic in the Global North: A Revival of Organic Ethics for a More Sustainable Food System

Milena Klimek, Valentin Fiala,
Bernhard Freyer, and Jim Bingen

Introduction

In this chapter we describe the growth of organic farming in the Global North from a grassroots, value-driven countermovement to a highly developed $72 billion market (Willer and Lernoud 2015). In doing so we discuss how the institutionalization, integration and professionalization that has accompanied the growth of the sector might endanger its ethical foundations. This discussion is important as organic farming is currently at a crossroads. On the one side there is the fear of a possible conventionalization of organic. Conventionalization is used to describe the fear that the current organic farming system has strayed from its original values and with this its potential to contribute to the transformation towards a more just and sustainable society (Darnhofer et al. 2009; Constance et al. 2015). On the other side the International Federation of Organic Agriculture Movements (IFOAM) is calling for an Organic 3.0 where organic farming and the organic movement, through its principles and practices, should further aid in the global development of mainstream sustainable food production (see "Organic Crossroads" section). In this fashion the positive characteristics of organic farming would be ideally shared with the broader food system—from large retail and discount stores to farmers' markets (FMs)—in order to affect more sustainable measures. We argue that the initial ethical foundations of organic farming as described in the IFOAM principles (see Table 7.2; see also IFOAM 2014) are relevant for the future of organics and that the values connecting them still play a large role in initiatives and developments that can drive agriculture towards more sustainable ways. To help illustrate this last point we present a comparative case study, in which we examine US and Austrian FMs and show the importance of values connected to organic as well as the general importance of ethical perspectives on these issues.

The Origins of Organic Values

In this section we briefly visit the beginnings of organic occurring in the Global North including the initial pioneers and the values they helped to embed in the movement. We later state that these values help illustrate how the ethics and motives of modern organic farming are different and are continuing to change from their origins. We argue that as the organic movement is at a crossroads it may be beneficial for its future to revisit them. The section after this covers these more modern organic developments in the United States and Europe roughly from the 1970s on.

As the historical roots of organic are shared here we go beyond organic as a way of farming towards organic as a movement. Organic farming originally emerged as several countermovements in opposition to the then beginnings of the industrialization of agriculture. Although differing from one another, these organic movements were kept together by comparable shared values resulting from a strong ethical orientation. This foundation of values is frequently used to question the movement's trajectory and often helps to maintain its cohesion.

The pioneers of the organic movement were first recognized in the early twentieth century, and were simultaneously yet independently developed differently throughout European countries and elsewhere (Reed 2010)[1] with what we know now to be values and practices of the modern organic movement. Between the two World Wars the movement began with different scientific and practice-oriented ideas. After World War II, the widespread production and commercialization of synthetic pesticides and fertilizers fostered organic farming's becoming of a more organized and well-known farming system (Vogt 2007). The movement especially gained momentum in the United States, through environmentalism sparked by Rachael Carson's scientific criticism on health related to DDT (Kinkela 2011). Table 7.1 presents some of the most influential organic pioneers of the Global North, their values and goals.

These pioneers not only contributed scientific or practical implications for organic farming but also offered the foundations of the organic farming values such as social aspects concerning human health, lifestyle, policy, ethical and spiritually related values, independence in trade, and relationships of farmers to consumers. These values can be considered as core aspects of the organic movement connected with the pioneers. They were integrated into the IFOAM principles as a result of the overall growth of organic farming and the recognition of shared values from the individual organic movements. In 1972, IFOAM was created as a unifying umbrella organization of diverse organic movements spread around the world. Here the original values have been revisited and used as a resource for the current organic IFOAM principles that also encompass more modern issues, and represent the international organic movement of today (see Table 7.2; Luttikholt 2007).

Development of the Organic Movement in the Global North

Since its origins organic farming has witnessed a strong increase in both land farmed and products purchased. Agricultural land under organic management increased globally from 11 million hectares in 1999 to 43.1 million or 1 percent of global agricultural in 2013. Simultaneously, the global organic market increased from $15.2 billion to $72 billion. Currently there are about 2 million certified

TABLE 7.1 Organic pioneers of the global north and their contributions toward the organic movement and its underlying values

Pioneers	Description of Initial Contributions to Organic and Its Foundational Values
Ewald Könemann (1899–1976)—Germany	Father of natural farming; he proposed a lifestyle focused on social issues such as vegetarianism and subsistence farming, as well as highlighting organic material cycles to improve soil fertility in low-input systems including abstinence from animals, synthetic and mineral inputs.
Rudolf Steiner (1861–1925)—Austria	Creator of anthroposophy and biodynamic agriculture; he combined ethical, ecological, and spiritual foundations to envision a farm as a closed system—its own organism—supplemented with on-farm organic fertilizers (see Steiner 1924). He envisioned a "dreigliederung" (three-way division) of society: (i) economy based on fraternity; (ii) law based on equality; and (iii) spiritual life based on freedom.
Sir Albert Howard (1873–1947) and Lady Eve Balfour (1898–1990)—England	Both focused on soil ecology and the relationships between plants, animals, soil and human health. Together, they are seen as the precursors of The Soil Association, still a large part of the UK's organic system today, and very much a frontrunner for soil health as the basis for overall health and "good" food. The soil association closely follows the current IFOAM principles.
Dr. Maria Müller (1894–1969) and Dr. Hans Müller (1891–1988)—Switzerland	Forbearers of organic-biologic agriculture, concentrated on soil fertility and the role of soil health in relation to the general health of a farm. Another important focus they had was on "bäuerliche Lebensweise," which focused on (political) independence, and the taking over of social responsibility for family farmers. This was embedded in a Christian ethic, which focused on stewardship of the divine creation.
Lord Walter James Northbourne (1896–1982)—England	A British agriculturalist closely following the biodynamic movement who first coined the term "organic farming." In his book *Look to the Land*, published in 1940, he highlighted the differences of farming at the time between organic farming and chemical farming, claiming the division of perspectives may potentially represent a lasting split in future worldviews.
Jerome Irving Rodale (1898–1971)—United States	An organic farming pioneer whose role was not so much of contributing new values, but furthering existing organic values. He did so specifically within the United States through publishing related articles, magazines, and books, thus publicizing the ecological and health values of organic farming and the movement itself. Eventually the Rodale Institute was opened, which specializes today in organic research, innovation, and outreach.

Source: Vogt 2007; Paull 2014; Freyer and Bingen 2015.

TABLE 7.2 IFOAM organic principles

Principle of health
Organic agriculture should sustain and enhance the health of soil, plant, animal, human, and planet as one and indivisible.
Principle of ecology
Organic agriculture should be based on living ecological systems and cycles, work with them, emulate them, and help sustain them.
Principle of fairness
Organic agriculture should build on relationships that ensure fairness with regard to the common environment and life opportunities.
Principle of care
Organic agriculture should be managed in a precautionary and responsible manner to protect the health and well-being of current and future generations and the environment.
Source: IFOAM 2009.

organic producers across the globe—258,000 in the EU and 12,880 in the United States (Willer and Lernoud 2015).

This growth transformed organic farming from social countermovements to an established part of the current western agrofood system in which many characteristics of organic have been changed. This has triggered a debate concerning the conventionalization of organic and questioning if this new type of organic is still sympathetic with the original values of the pioneers and if organic is still a catalyst for transformation towards a more sustainable and just society.[2] The diversity of changes within the organic movement can be summarized in three main developments: Institutionalization, Integration, and Professionalization.

In the following we briefly present these developments. We focus on the situation within the EU and the United States because these two regions hold the largest market for organic products. In 2013, 83 percent of the global sales of organic products occurred in these two regions (Willer and Lernoud 2015). Furthermore the origins of organic farming in Europe and the United States are often seen as influential for the global organic movement (Willer and Meredith 2015).

Institutionalization

The institutionalization of organic farming is one of the most striking developments of the sector (see Michelsen et al. 2001). From its origins a standardized "organic," controlled by a three-party certification regime, slowly emerged in the Global North and spread around the world (Hatanaka et al. 2005; Fouilleux and Loconto 2016). Early on, organic faced strong opposition from established agricultural institutions (Jurtschitsch 2010; Uekötter 2012; Willer 2016) and the organic sector, up until the 1980s, remained small and had few connections with other agricultural institutions. Prior to institutionalization, organic standards were negotiated within smaller farmer groups and not guaranteed via certification but via mutual trust between consumers and producers created through direct marketing (or very short supply chains) (Pollan 2006; Freyer and Bingen 2015). With the ongoing growth of organic, the organic community created more strict and harmonized rules for its members. In addition—especially in Europe—organic farmers established connections within existing agricultural institutions looking for support and the recognition of their way of farming (Michelsen et al. 2001; Moschitz et al. 2004).

In Europe national governments began, in the mid- to late 1980s, to view organic as a means to challenge the problem of overproduction or to support farmers working in difficult conditions (e.g., mountainous regions; Willer 2016). These actions led to the establishment of the first EU-wide regulation of organic production by the EU-Commission (Vogl and Axmann 2016). This was a crucial moment for organic. Suddenly organic farming was defined through—and, in order to label products as organic, required to abide by—an official "checklist" of standardized rules. To ensure this, a strong third party certification was established. Thus the regulatory power over organic was transferred from the organic community to the public sector and private certification bodies, in which the EU controls the standards and essentially defines organic (Sanders 2016).

This way of defining and controlling organic farming became the role model of organic around the world. In 2013, eighty-two countries had comparable standards (Willer and Lernoud 2015). This legal definition of organic farming was not only a precondition for the integration of organic in the mainstream market and the allocation of price premiums (see next section), but also for the possibility to subsidize organic farming with public money.

In the United States the creation of a national organic standard took longer. Private standards and a connected private certification and control system were established in the early 1970s (Kuepper 2010). The Alar scandal in 1989,[3] the subsequent increase in demand for organic apples, and the connected growth of labeling resulted in the organic food production act of 1990 and the first public definition of organic. However, it took twelve years for a national standard of organic products in the United States to be implemented (Kuepper 2010). In contrast to Europe, the United States did not prefer subsidies but market mechanisms to foster the growth of organic (Constance et al. 2015).

Integration

The standardization and regulation of organic farming were important reasons for the boom of organic farming in the 1990s[4] because it enabled the integration of organic into the mainstream—some would say conventional—supply chains. It helped to bring organic out of specialty stores into the supermarkets and therefore to a wide range of consumers.

Institutionalization as a development in organic was seen by many as a necessary step in this change because it is difficult to check an organic product's organic qualities, even after purchase. Such products are known as "credence products." Thus ideal organic regulation extends the idea of trust for the consumer (Yiridoe et al. 2005). The sale of organic products is essentially based on trust. Prior to integration, organic farming in the EU and United States was clearly a niche market with specific target groups and short and direct marketing channels.

Here direct social relations created the necessary trust. Modern mainstream food supply chains are long, complex, and create distance between producers and consumers (for an overview, see Goodman and Watts 1997). Therefore other means to create trust are necessary and the establishment of clear standards and an independent control system are two of the most important ones.[5]

The integration in mainstream supply chains altered the characteristics of the Western organic agrofood system dramatically. Due to the rapid change of marketing channel distribution, the current organic retail sector in the United States bears almost no resemblance to the situation prior to the 1990s (Dimitri and Oberholtzer 2009). In 1991 only 7 percent of organic sales were conducted at conventional retailers, In 2006 that number had jumped to 46 percent (Dimitri and Oberholtzer 2009)[6] and in 2010 mass-market grocery stores, such as Walmart and Krogers, alone accounted for 54 percent of organic food sales (Constance et al. 2015). In the EU there is a similar situation; conventional retailers dominate the organic food supply chain (Willer and Lernoud 2015).

Finally, it has to be mentioned that organic integration into mainstream supply chains is not a phenomena restricted to the Global North, but has worldwide effects. As we mentioned, the United States and Europe combined is by far the largest market for organic products (83 percent of global organic sales in 2013) and although organic production in other areas is increasing, consumption of organic in such areas remains low (Willer and Lernoud 2015). Therefore organic supply chains have become global, international trade relations are increasing in importance, and producers in the global south supply consumers in the Global North.

Professionalization

The third significant development—professionalization of organic farming—is highly intertwined with the two previous developments and encompasses several aspects. The first is the establishment of professional organic producer associations. Early organizations connected with organic farming in Europe were social movements, which did not specifically represent farmer interests. It was not until later that professional farmer associations were formed and grew in their power and influence. In 1972 the IFOAM was founded. Acting as an international umbrella organization for organic farmers by representing their interests, IFOAM formulated principles and standards for organic farming that act as a guideline but have no legal basis. In 2014 IFOAM was affiliated with 812 organizations in 120 countries. Fifty-one of them were in the United States with the majority in the EU (Willer and Lernoud 2015).

The second aspect of professionalization is the organization of research and development of organic farming as well as connected education and training. Originally organic education and training was mostly completed within the movements and organized by farmers themselves. Education and training of organic in the EU was partially taken over by other agricultural organizations. For Switzerland Niggli (2002) identified four phases of organic research and development: (i) pioneering work of farmers, (ii) establishment of private research institutions, (iii) first chairs at universities, and (iv) the acceptance of this research field within publicly funded research. This development generally fits that of the Global North, and—although private farmer associations still play a large role in this regard—research and education has been directed away from farmer associations to other institutions.

The third aspect of professionalization, the possible professionalization of the farmers themselves, is perhaps not as straightforward as the other aspects and is likewise debated. Naturally, early organic farmers conducted themselves professionally. However here we highlight organic farming becoming a *profession*. Since the beginning of organic a change in lifestyle has been promoted by the movement including diet and being critical of the wider agrofood system (Freyer and Bingen 2015). With its growth and connected changes organic farming became a new profession among agriculture. Thus, although contested,[7] there is research that suggests that motives for converting towards and practicing organic farming and therefore the values of actors have become increasingly utilitarian and economically oriented (Michelsen et al. 2001; Padel 2001; Michelsen and Rasmussen 2003; Läpple and Van Rensburg 2011).

Organic Crossroads: Conventionalization or Organic 3.0?

The developments described above have triggered an extensive debate about the possible conventionalization of organic.[8] This debate continues today as there has been no consensus concerning the widespread existence of conventionalization of organic farming (see Darnhofer et al. 2009). Yet this debate also illustrates that many organic actors believe that the organic movement is at an important developmental turning point. IFOAM itself took the initiative recently to publish a strategy to drive organic development to the next level: Organic 3.0.[9] Envisioning an organic movement breaking out of its 1 percent niche, Organic 3.0 is hoped to become the role model for a global sustainable agriculture. To achieve this, Organic 3.0 aims to move beyond its currently restrictive and controlling certification system, seen as demanding only minimum standards and therefore working as an end-of-pipe solution. Conversely, the focus should be a system that ensures continuous improvement. Ideally to increase the sustainability of the overall agrofood system, Organic 3.0 should work more closely with other sustainable initiatives and conventional farmers. The Organic 3.0 strategy consists of six points (for more details see Arbenz et al. 2015):

1. Establishing a **culture of innovation** along the food chain while balancing the impacts of technology

2. **Continuous improvement towards best practice**

3. **Diverse ways to ensure transparent integrity** beyond third party certification

4. **Inclusion of wider sustainability interests** through alliances with other initiatives

5. **Holistic empowerment from the farm to the final product**

6. And realizing **true value and fair pricing**—true cost accounting and internalizing external effects.

In this chapter we do not wish to argue that Organic 3.0 document is the be-all and end-all of the organic movement; however, the idea is gaining momentum and is attempting to openly address significant issues within the movement. Important to remember is that Organic 3.0 is a part of a discussion that critically questions the past development of organic and brings its ethical foundation to the forefront of public attention. It reminds us that there are other possible development pathways than the one that leads to organic's current trajectory.

Thus, organic is at a crossroads. On the one side, "organic business as usual" contains the threat of possible conventionalization endangering the foundation of the organic movement. On the other side, a pathway exists that is created through reflection of the organic status quo pertaining to its initial values and ethical foundations—in which Organic 3.0 is one example of this process. Therefore, we argue that it is essential to revisit and emphasize the ethical foundations of organic. Although these foundations are often overlooked we believe they are a key aspect in choosing the appropriate future direction of organic. Keeping the foundations as a focus in the organic movement may aid in avoiding or meliorating conventionalization and may truly enable an Organic 3.0 as envisioned by IFOAM. In the following we present a short case study that illustrates the importance of organic values in the creation of sustainable agrofood systems.

Farmers' Markets: A Case Study on the Importance of Values for the Future of Organic

In the following we wish to use a practical example to illustrate our point that the ethical foundation of organic is essential for its future, and for a future called Organic 3.0. To do this we compare two different types of FMs in the Global North (in Minneapolis, Minnesota, and in Vienna, Austria; see Klimek et al. 2018). Both types of FMs have their place in the modern agrofood system and they are often seen as the most prevalent and successful venues in sustainable food systems, specifically for small producers.

One focus of the Klimek et al. (2018) study was to understand what role organic values play in the way FMs function. It revealed IFOAM principle–related values present in the different FMs; how these values were expressed through rules and norms and what their role was to the functioning of the FMs. For this chapter, we focus on how values found in the FMs of this study are associated with the organic movement's values (see Tables 7.2 and 7.3). We show that the values present in FMs influence FM functions and enable them to play a role for the further development of organic farming.

In examining values at FMs it is important to first understand that FMs are not solely economic institutions, but spaces where economy is socially embedded (Hinrichs 2000).[10] Therefore economy is rooted in the noneconomic values of a FM, such as trust between farmers and consumers or food and farming related educational events sponsored by the markets themselves. Such values are created and recreated and revealed through FM norms and rules. They are essential for the functioning of a FM and are part of actively shaping food supply chains.

To determine how values were expressed in the form of rules and norms in six Viennese and six Minneapolis FMs, a value-based framework was designed. The framework integrated IFOAM

FIGURE 7.3 *Farmers' market poultry. Photo courtesy of Janet Chrzan.*

organic principles, aspects from Generative Economy (Kelly 2012), and FM elements deemed necessary for FM success (Stephenson 2008). It also conceptually organized forty-two qualitative interviews with market managers and farmers/vendors as well as Dot Surveys[11] conducted at the Austrian FMs. Participatory observation, in the form of both volunteering and selling at the FMs as well as being a consumer, was also completed at each market. These methods were used to reach all three main FM stakeholder groups—market managers, farmers/vendors, and consumers. Involving all major stakeholder groups allows for FMs to be examined as systems that include the interactions and relationships involved in the physical space in which the FM takes place.

The cross-national element of the study enabled the illustration of the differences in values present in two varying metropolitan FM regions. To show one example, Tables 7.3 and 7.4 describe the differences in the FM regions using the values expressed in the Principle of Care (see Table 7.2). The results clearly indicate that the sheer amount of IFOAM-related values expressed in the Minneapolis FM norms and rules were far more than those at the Viennese FMs *and* the values found were more likely to be shared by all FM stakeholder groups in Minneapolis as opposed to the FM groups of Vienna.

This diversity is seen mainly because Minneapolis FMs are structured and governed in a way that allows for values to be shared between the management, vendor/farmer and consumer stakeholder levels. Sometimes the reason why these FMs are governed in such away is because of the values the FMs were originally based upon. Therefore, in Minneapolis FMs, farmer support

TABLE 7.3 Value illustration: Minneapolis

Examples of values represented as the *principle of care* in the norms and rules of Minneapolis FMs	
Value*	Examples
Importance of education	• Educational programs for the public: 　◦ Community booths (a rotating booth inviting local organizations to educate and make themselves known from biking and yoga groups to master gardeners and local composters) 　◦ Weekly-monthly cooking demonstrations 　◦ Updates on food and farming policies and regulations 　◦ Children/family friendly programs • Educational programs for vendors are also available: 　◦ How to aesthetically present their products 　◦ Rules and regulations of homemade products
Support of small farmers	• The impetus of most markets was to provide access to healthy, fresh, and local food to a specific neighborhood; the best way to achieve this was seen as going directly to farmers • Each FM has a designated percent of actual farmers that must be selling or present at the markets • There is an ongoing discussion at all FMs of how to broach the true value of food topic • Education concerning farming issues and policies
Support of FM alterity	• This occurs through the community built around direct marketing and a friendly and engaging atmosphere • Many issues related to specific FM values are promoted, for example, sustainability in hosting community days to encourage bike riding, composting, starting a garden, etc.
Sustainable economics	• Enables direct marketing • Current ongoing discussion at all FMs of how to tackle the true value of food—for example, what is involved in the production of quality food and how that might be reflected in price (fair prices for consumers vs. fair wages for farmers and laborers).
Community development	Community is defined by the markets as being more than consumers and the surrounding neighborhood but the farmers and their farms, thus building community comes in the following forms: 　◦ Support of other local businesses 　◦ Support of the FMs as a different kind of economy 　◦ Goal of connecting people to their food and food producers 　◦ Support of local and small farmers 　◦ Cooperation among vendors (highlighting products that go well together, etc.) 　◦ Reduced competition among vendors 　◦ Foreign language and communication sensitivity 　◦ Inclusion of community / regional food traditions 　◦ Education

Examples of values represented as the *principle of care* in the norms and rules of Minneapolis FMs	
Animal treatment	• Many farmers bring pictures of their farm animals and farming conditions • Some markets conduct farm visits to ensure transparency and production practices • Some farmers are organic and comply with strict animal production standards
Cooperation	• Many partnerships are seen between vendors • Most of the markets depend on volunteers • Extended FM community members such as local businesses, organizations, and restaurant chefs often make partnerships with FM members or share their knowledge there
Reduced competition	• With the exception of one FM, farmers and vendors are chosen based on needs of the market to broaden diversity of product for consumers and reduce competition between vendors
Tradition	• Highlighting traditional Minnesota products (e.g., wild rice, maple syrup, mushrooms, honey, etc.) as well as traditional cultures (Native American) • Most FMs or vendors at some point highlight the "romanticism" of farm life reaching to themes often associated with "the past" (e.g., craft products, handmade products, size, scale, family, etc.). This often occurs either in their production practices or marketing

Farmer involvement	• All FMs have at least one annual (up to monthly) meeting for vendor input & information
	• Farmers are often part of the decision making process in the FMs:
	  ◦ Many FMs have a farmer advisory board to help managers understand issues and concerns
	  ◦ One FM is run by a growers association, where board members are farmers

*These values were found inductively and were not predefined or exclusive. The same values are used here as those in the Minneapolis case to illustrate a rich comparison of both FMs. You will see that many values found in Minneapolis are not expressed in the norms and rules of Vienna's FMs.
Source: Drawn from Klimek et al. (2018).

occurs in the attraction of more consumers to the market while simultaneously supporting returning consumers in offering additional values to the tacit knowledge and products farmers provide such as music, children's programs, and education in food and farming related topics—for example, cooking, health, gardening, composting, biking, etc. In viewing the Minneapolis FMs through the organic principles we see how organic values can be shared, reproduced, and benefit FMs in Minneapolis and other metropolitan FMs.

All of the Viennese FMs, on the other hand, are governed municipally and are perceived by the city as neoclassical markets—not as alternatives but in the same class as supermarkets. The way these FMs are governed, based on values focused primarily on food access, safety, and hygiene, creates difficulty in ensuring that values are shared between all stakeholders. The municipal governance structure often inhibits regular marketing strategies and educational opportunities that support both consumer and farmer values (see Table 7.4 for examples of these obstacles). This detachment among stakeholder values creates extra work for vendors/farmers in expressing the values embedded within both their products for sale and the more contemporary perspectives of what a FM might be.[12]

TABLE 7.4 Value illustration: Vienna

Examples of the values represented as the *principle of care* in the norms and rules of Viennese FMs	
Value*	Examples
Education	• There are a few cooking demonstrations giving each year in a few of the markets or individual events
Support of small farmers	• There is not a priority of having farmers from the management side of the FM; farmers as vendors are, however, given priority over resellers for stand space (if interest exists)
Support of FM alterity	• FMs are managed as if they are still a main source for food procurement in Vienna, in competition with supermarkets
Sustainable economics	• FMs are managed from a neoclassical perspective
Community development	• Due to the lack of educational programs, community is built directly from the relationships between farmers/vendors and consumers
Animal treatment	• Many vendors are organic and therefore production partakes in strict certification and regulation
Cooperation	• Very few examples of farmers and vendor networks and partnerships were found
Reduced competition	• The FM system in Vienna is all managed through a lottery process, this means that there can be and are markets that have 80–90 percent of their stands filled with vegetables and fruit (no diversity = disappointed costumers)
Tradition	• There is a small Slow Foods presence that showcases traditional foods and ways of production • A few farmers highlight rare and traditional plant varieties and breeds, and focus on traditional foods
Farmer involvement	• There are no meetings held specifically for the FMs nor are farmers included in decision making processes

*These values were found inductively and were not predefined or exclusive. The same values are used here as those in the Minneapolis case to illustrate a rich comparison of both FMs. You will see that many values found in Minneapolis are not expressed in the norms and rules of Vienna's FMs.
Source: Drawn from Klimek et al. (2018).

FMs encompass the entire food value chain from farm to fork. This enables the expression of a wide variety of values connected to the IFOAM ethical principles. A strong presence of values related to the ethical foundation of the organic movement creates potential for FMs to contribute to the strategy points of Organic 3.0 mentioned above. Examples of these contributions are explained in Table 7.5 using the example case of the Minneapolis FMs.

FIGURE 7.4 *Our chickens don't do drugs! Photo courtesy of Janet Chrzan.*

The value sets of FMs in Minneapolis encourage and instigate face-to-face contact, creation of trust, short supply chains, and attempt to bridge the disconnect within the current food system through education and atmosphere. These values to actions enable the FM movement in the United States to work within the goals of Organic 3.0 today. Thus, because of the prevalent presence of IFOAM values in US FMs they contribute more to Organic 3.0 than do the Vienna FMs.

In this section we indicated that values are essential for the functioning of a FM, are also intrinsic to the way they function as a whole, and are part of actively shaping food supply chains. Specifically, we showed here that because FMs in Minneapolis follow values related to the IFOAM principles those FMs (or other FMs or parts of the agro foodsystem that share such values) play a more important part for the future of organic and even more so in a future that chooses the Organic 3.0 pathway, than do FMs—such as those in Vienna—that do not share such values.

Conclusion

Despite the changing ethics and motives of modern organic farming through institutionalization, integration, and professionalization we have argued here that the initial ethical foundations of the organic pioneers, now described in the IFOAM principles, are still crucial for the future development

TABLE 7.5 Potential contributions of Minneapolis FM values to the six strategies of Organic 3.0

Organic 3.0 Strategy	Potential Contributions
Culture of innovation	• An FM's community and atmosphere strengthens its values, thus encouraging their reproduction—for example, some FM vendors may see how well organic producers are received at the FMs and decide to convert (important to 3.0) • As socially embedded places of economy, FMs combine social, ecological and technological innovations (important to 3.0) from production to marketing levels
Continuous improvement towards best practice	• Direct face-to-face contact between consumers and farmers produces an element of trust in the FMs; this paired with many FM initiatives of transparency through farm visits furthers best practice based on morals and values not necessarily through certification
Diverse ways to ensure transparent integrity	• Highlighting specific farms and farming practices during the FMs • Managers conducting farm visits • "Open house" days for customers to visit farms • Trust built through direct marketing • Organic certifications

Inclusion of wider sustainability interests	• Community booths hosting themes of:
	  ◦ Gardening
	  ◦ Composting
	  ◦ Personal health—for example, yoga at the market and biking to the market days
	  ◦ Native to MN day—highlighting indigenous communities and related FM products

Holistic empowerment from farm to final product	• FMs encompass the entire food/value chain from production to consumer and has the opportunity to encompass empowerment along this chain (this aspect seems to be the next step in FM practice and research as gender, race and economic bracket of typical FM shopper is becoming a hot topic) some current examples are: • Including farmers in the FM decision making processes • Market 101 classes in various languages for market customers lacking knowledge about certain products or how to cook them • Food accessibility through programs creating awareness around electronic benefit transfer (formerly the food stamp program) at the FMs
True value and fair pricing	• Although this theme was just beginning to be addressed at the FMs during the time of data collection one example is: ◦ During market hours showing price comparisons of market prices at natural food coops (the main competition for FMs) and at FMs.

Source: Drawn from Klimek et al. 2018.

FIGURE 7.5 *Farmers' market bounty. Photo courtesy of Budd Cohen.*

of the movement. Through the metropolitan FM case study, we have illustrated how the organic principles and the values connected with them can be both integrated into the current agrofood system and play a large role in initiatives, specifically in the Global North, that can push food and farming related agendas together towards more sustainable ways. Therefore our main argument should now be better understood: that in the confusion of the current organic situation and its desired future, it is its ethical foundation that can help guide and evaluate the next steps taken to bring the movement further. Thus together, the IFOAM principles create the necessary signpost at the crossroads of which the organic movement is currently facing.

NOTES

1 Parallel to these organic origins influential in the United States and EU, other forms of organic and sustainable farming methods were emerging around the world (e.g., Efraim Hernandez Xolocotzi in Mexico or Masanobu Fukuoka in Japan; Arbenz et al. 2015).

2 For an accurate and current overview of the conventionalization debate, see Darnhofer et al. 2009 or Constance et al. 2015.

3 Resulting in a public discussion about the negative health effects of the application of certain pesticides on apples.

4 For example, the number of organic producers within the EU increased from less than 20,000 in 1990 to 120,000 in 1999 (Willer and Yussefi 2000).

5 Another important attempt to create trust within mainstream supply chains is the introduction of independent organic labels and brands from retailers (Dimitri and Oberholtzer 2009).

6 This trend continues to increase: In 2012, 78 percent of US certified organic farms contributed to wholesale, 14 percent sold to retailers, and only 6 percent to consumers directly (National Agricultural Statistics Service 2015).

7 For example, in a later study, Padel (2008) conducted focus groups in five European countries (Austria, Italy, Netherlands, the UK, and Switzerland) and concluded that there are only small differences between the values of established farmers and newcomers.

8 Within the scientific community this debate began with the work of Buck et al. (1997) and gained importance with further research from Guthman (2004) concerning conventionalization of organic farming in California.

9 IFOAM describes Organic 1.0 as the time of the organic pioneers (comparable with the "Organic Origins" section above) and Organic 2.0 as the establishment of a regulated and certified organic agrofood system from the 1970s until today (comparable with the "Development of Organic" section above).

10 Diverging from neoclassical economics, economist Karl Polanyi's concept of social embeddedness has often been used to describe economic interactions at a farmers' market as being embedded in the social. Therefore, the significance of noneconomic elements in the way a system—here the farmers' markets—functions is taken into consideration (Polanyi 1944; Polanyi et al. 1971).

11 Dot Surveys are specifically designed to obtain data at FMs (see Lev et al. 2007). The surveys are interactive with usually four to five closed questions posed on flip charts at markets in which consumers receive "dot" stickers to signify their answers. Although limited in scope, the benefit of such a survey is positive interaction with consumers and a high number of participants.

12 FMs in the Global North are increasingly acknowledged as places of alterity, displaying the social embeddedness (see Hinrichs 2000; Feagan and Morris 2009) of values such as farmer support, trust, education, and atmosphere through the direct marketing venue (Kirwan 2004).

References

Arbenz, M., Gould, D., and Stopes, C. 2015. *Organic 3.0 or Truly Sustainable Farming and Consumption*. Bonn: IFOAM—Organics International.

Buck, D., Getz, C., and Guthman, J. 1997. From Farm to Table: The Organic Vegetable Commodity Chain of Northern California. *Sociologia ruralis* 37: 3–20.

Constance, D., Young, J. C., and Lara, D. 2015. Engaging the Organic Concentionalization Debate. In *Re-Thinking Organic Food and Farming in a Changing World*, ed. Bernhard Freyer and Jim Bingen. Dordrecht: Springer.

Darnhofer, I., Lindenthal, T., Bartel-Kratochvil, R., and Zollitsch, W. 2009. Conventionalisation of Organic Farming Practices: From Structural Criteria towards an Assessment Based on Organic Principles. A review. *Agronomy for Sustainable Development* 30: 67–81.

Dimitri, C., and Oberholtzer, L. 2009. Marketing U.S. Organic Foods—Recent Trends from Farms to Consumers. In *Economic Information Bulletin*. Washington, DC: US Department of Agriculture.

Feagan, R. B. and Morris, D. 2009. Consumer Quest for Embeddedness: A Case Study of the Brantford Farmers' Market. *International Journal of Consumer Studies* 8: 235–43.

Fouilleux, E., and Loconto, A. 2016. Voluntary Standards, Certification and Accreditation in the Global Organic Agriculture Field. A Tripartite Model of Techno-politics. *Agriculture and Human Values* 34: 1–14.

Freyer, B., and Bingen, J. 2015. *Re-thinking Organic Food and Farming in a changing World*. Dordrecht: Springer.

Goodman, D., and Watts, M. 1997. *Globalising Food: Agrarian Questions and Global Restructuring*. New York: Routledge.

Guthman, J. 2004. The Trouble with "Organic Lite" in California: A Rejoinder to the "Conventionalisation" Debate. *Sociologia ruralis* 44: 301–16.

Hatanaka, M., Bain, C. and Busch, L. 2005. Third-Party Certification in the Global Agrifood System. *Food Policy* 30: 354–69.

Hinrichs, C. C. 2000. Embeddedness and Local Food Systems: Notes on Two Types of Direct Agricultural Market. *Journal of Rural Studies* 16: 295–303.

IFOAM. 2009. *The Principles of Organic Agriculture*. www.ifoam.org/about_ifoam/principles/index.html (accessed January 9, 2011).

IFOAM. 2014. *Principles of Organic Agriculture*. www.ifoam.org/en/organic-landmarks/principles-organic-agriculture (accessed August 30, 2014).

Jurtschitsch, A. 2010. Bio-Pioniere in Österreich. *Vierundvierzig Leben im Dienste des biologischen Landbaus. Böhlau, Grüne Reihe des Lebensministeriums* 21.

Kelly, M. 2012. *Owning Our Future: The Emerging Ownership Revolution*. San Francisco: Berrett-Koehler.

Kinkela, D. 2011. *DDT and the American Century: Global Health, Environmental Politics, and the Pesticide That Changed the World*. Chapel Hill, NC: University of North Carolina Press.

Kirwan, J. 2004. Alternative Strategies in the UK Agro-Food System: Interrogating the Alterity of Farmers' Markets. *Sociologia ruralis* 44: 395–415.

Klimek, M., Bingen, J., and Freyer, B. 2018. Metropolitan Farmers Markets in Minneapolis and Vienna: A Values-based Comparison. Agriculture and Human Values. *Agriculture and Human Values* 35: 83–97.

Kuepper, G. 2010. *A Brief Overview of the History and Philosophy of Organic Agriculture*. Poteau, OK: Kerr Center for Sustainable Agriculture.

Läpple, D., and Van Rensburg, T. 2011. Adoption of Organic Farming: Are There Differences between Early and Late Adoption? *Ecological Economics* 70: 1406–14.

Lev, L., Stephenson, G. O., and Brewer, L. 2007. Practical Research Methods to Enhance Farmers' Markets. In *Remaking the North American Food System—Strategies for Sustainability*, ed. Clare Hinrichs and Thomas Lyson. Lincoln and London: University of Nebraska Press.

Luttikholt, L. W. M. 2007. Principles of Organic Agriculture as Formulated by the International Federation of Organic Agriculture Movements. *NJAS-Wageningen Journal of Life Sciences* 54: 347–60.

Michelsen, J., Lynggaard, K., Padel, S., and Foster, C. 2001. Organic Farming Development and Agricultural Institutions in Europe: A Study of Six Countries. In *Organic Farming in*

Europe: Economics and Policy, ed. Stephan Dabbert. Stuttgart–Hohenheim: Universität Hohenheim.

Michelsen, J., and Rasmussen, H. B. 2003. Nyomlagte danske økologiske jordbrugere 1998. En beskrivelse baseret på en spørgeskemaundersøgelse.

Moschitz, H., Stolze, M., and Michelsen, J. 2004. Report on the Development of Political Institutions Involved in Policy Elaborations in Organic Farming for Selected European States. In *Further Development of Organic Farming Policy in Europe with Particular Emphasis on EU Enlargement*. Odense: University of Southern Denmark.

National Agricultural Statistics Service. 2015. Organic Survey. In *2012 Census of Agriculture*, ed. Joseph Reilly. Washington, DC: US Department of Agriculture.

Niggli, U. 2002. The Contribution of Research to the Development of Organic Farming in Europe. Paper read at UK Organic Research 2002: Proceedings of the COR Conference.

Padel, S. 2001. Conversion to Organic Farming: A Typical Example of the Diffusion of an Innovation? *Sociologia ruralis* 41: 40–61.

Padel, S. 2008. Values of Organic Producers Converting at Different Times: Results of a Focus Group Study in Five European Countries. *International Journal of Agricultural Resources, Governance and Ecology* 7: 63–77.

Paull, J. 2014. Lord Northbourne, the Man Who Invented Organic Farming, A Biography. *Journal of Organic Systems* 9: 31–53.

Polanyi, K. 1944. *The Great Transformation: The Political and Economic Origins of Our Time*. Boston: Beacon Press.

Polanyi, K., Arensberg, C., and Pearson, H. W. 1971. *Trade and Market in the Early Empires: Economies in History and Theory*. Chicago: Henry Regnery.

Pollan, M. 2006. *The Omnivore's Dilemma: A Natural History of Four Meals*. London: Bloomsbury.

Reed, M. 2010. *Rebels for the Soil: The Rise of the Global Organic Food and Farming Movement*. New York: Routledge.

Sanders, J. 2016. Agrarpolitik. In *Ökologischer Landbau*, ed. Bernhard Freyer. Bern: Hauptverlag.

Steiner, R. 1924. *Geisteswissenschaftliche Grundlagen zum Gedeihen der Landwirtschaft*. Landwirtschaftlicher Kurs. Stuttgart: Steiner Verlag.

Stephenson, G. O. 2008. *Farmers' Markets: Success, Failure, and Management Ecology*. Amherst, NY: Cambria Press.

Uekötter, F. 2012. *Die Wahrheit ist auf dem Feld: eine Wissensgeschichte der deutschen Landwirtschaft*. Göttingen: Vandenhoeck & Ruprecht.

Vogl, C. R. and Axmann, P. 2016. Regelungsmechanismen. In *Ökologischer Landbau*, ed. Bernhard Freyer. Bern: Hauptverlag.

Vogt, G. 2007. The Origins of Organic Farming. In *Organic Farming: An International History*, ed. William Lockeretz, 9–29. Wallingford: CABI.

Willer, H. 2016. Geschichte. In *Ökologischer Landbau*, ed. Bernhard Freyer. Bern: Hauptverlag.

Willer, H., and Lernoud, J. 2015. *The World of Organic Agriculture. Statistics and Emerging Trends 2015, FIBL-IFOAM Report*. Frick and Bonn: FIBL and IFAOM—Organics International.

Willer, H., and Meredith, S. 2015. Organic Farming in Europe. In *The World of Organic Agriculture. Statistics and Emerging Trends*, ed. Helga Willer and Julia Lernoud. Frick and Bonn: FIBL and IFOAM—Organics International.

Willer, H., and Yussefi, M. 2000. *Organic Agriculture Worldwide—Statistics and Future Prospects*. Frick and Bonn: FIBL and IFOAM.

Yiridoe, E. K., Bonti-Ankomah, S., and Martin, R. C. 2005. Comparison of Consumer Perceptions and Preference toward Organic versus Conventionally Produced Foods: A Review and Update of the Literature. *Renewable Agriculture and Food Systems* 20: 193–205.

8

Profile: Erika Tapp, Concerned Consumer

As a resident of Philadelphia and someone whose professional work has focused on city-based initiatives, I really appreciated the focus on the small-scale efforts of individual cities. Too often, I think we expect macro-level change to come from federal or international agencies. Programs created and driven by local municipalities are an excellent space to test new models, and ideally, show that it is economically possible (or even beneficial) to be environmentally sustainable. The cases of Lejre and Copenhagen presented by Kristensen and Hansen reflect the many and varied approaches cities are taking to encourage healthy eating, developing a market for organic produce, and generally becoming more sustainable. Both resonated with me and my experiences as a gardener and resident in the City.

In Philadelphia, we have seen both bottom-up and top-down approaches to becoming sustainable. And like both of these cases, have seen varying degrees of success. Though, across the board, clarity of vision, collective buy-in, integration of efforts, and leadership seem to be hallmarks of success.

Like Lejre, Philadelphia has a number of grassroots or community-driven projects to improve the availability of fresh, local food as well as to create a greener and cleaner city. Restaurateurs are establishing urban farms in underserved neighborhoods and provide community garden space, and programming for kids. They then use the produce (almost exclusively organic) on their menus. This model of hyper-local sourcing is mutually beneficial for the communities and the restaurants.

There have been dozens of community gardens started by neighbors, rich and poor alike. Their interest is in growing their own food. They are frequently located on unsecured land (i.e., is privately held, ability to use as gardening space is not guaranteed year-to-year, or ownership is unknown), so may be bulldozed at any moment. Typically, individuals are given plots to maintain, and all are expected to participate in community workdays. This has not proven to be a recipe for long-term access to organic, local produce. It has decidedly improved a sense of community among gardeners and near-neighbors. These gardens are located on otherwise vacant land that would otherwise be an eyesore or a safety liability.

The Pennsylvania Horticultural Society (PHS), an independent nonprofit that holds a number of contracts with the city, has embraced the explosion of interest in community gardens. They have a number of training programs, both in the technical aspects of gardening as well as some of the

FIGURE 8.1 *Erika Tapp at the farmers' market. Photo courtesy of Janet Chrzan.*

brass tacks of actually managing one (securing the land, water sources, organizing volunteer days, etc.). PHS also has a program called City Harvest, wherein inmates in the Philadelphia Prison System start seedlings that are then distributed to gardens throughout the city. Those gardens then donate the produce to food cupboards and other neighborhood-based food distribution systems. This is a great example of a singular program having multiple benefits—prisoners learn gardening and landscaping skills, gardens receive free vegetable starts, and, at least some, food insecure neighbors have access to free, hyper-local produce. Like Lejre, coordination of grassroots and governmental efforts has resulted in a positive program achieving multiple goals.

FIGURE 8.2 *Erika in the garden. Photo courtesy of Janet Chrzan.*

Consumers, Citizens, and the Participatory Processes on Organic Food: Two Case Studies from Denmark

Niels Heine Kristensen and Mette Weinreich Hansen

Introduction

Global, national, and local actors are continuously influencing food systems and food chains. Local food economies and structures have for many decades been strongly influenced by global regimes without leaving much room for maneuver for local actors. Locally embedded interests in sustainable and organic food economies and networks have been gaining greater access to regional and national urban market channels. These are often challenged by weaker access and applicability to urban market channels (retail chains), be it based on scale, information technology, or the like.

Reforms of public governance structures and the redesign of structures of municipalities and local governments have encouraged the application of New Public Management theories and principles (Devas and Delay 2006). This has, in some Danish municipalities, offered progressive actors and decision-makers opportunities for creating new policies and new partnerships.

The European Union installed its first voluntary regulation on organic products in 1991. This regulation included plant production standards and was followed by labeling and regulation standards for organic animal products in 1999. In 1987 Denmark was the first nation to pass a bill that defined standards for organic farming. This was followed by several national action plans, the last one being national action plan 2020 (Ministry of Environment and Food 2012).

Before this public regulation was implemented, organic farming and organic food were regulated by private organizations and through social and market networks. In particular, the member-owned retailer "COOP" drove the marketing and sales of organic products in cooperation with organic pioneers, organic farmers and organic farmer associations. Since then, the organic market in Denmark has been increasing successfully because organic produce has been available in ordinary supermarkets and in discount supermarkets, supported by very high trust in—and acquaintance with—the state regulated organic label.

To increase organic production and reverse an observed stagnation in the conversion of organic farmland, the former Social-democrat government launched an Organic Vision 2020 (Ministry of Environment and Food 2012). Public kitchens were seen as key actors to boost demand for organic products, which would stimulate conversion to organic farmland. Thus there was in this vision an expected direct link between public food and organic farmland since the demand side (kitchens) would result in a need for more farmers (suppliers). In 2012, to further this vision, a national support scheme for transition of public kitchens to use organic food was implemented. This was the result of a parliament decision to use public food procurement to drive the conversion from conventional to organic farming through market development. The aim was to encourage use of at least 60 percent organic products measured in turnover (in either kroner or kilogram) with a menu-labeling scheme ranking the level of organic products as either gold, silver or bronze (Ministry of Food and Environment 2012). This support strategy supported education and competence building of kitchen staff and several municipalities, regional and state kitchens took part.

The implementation of national policies for organic conversion has led to more positive cooperation between public and private actor groups. This strategy of developing strong relations between private and public actors and institutions resulted in a stronger embedding of organic practices, procedures and provisioning and has resulted in better knowledge about the differences of conventional and organic farming principles among the population. Organic farming today covers 7.5 percent of the agricultural area in Denmark and the organic products share of the daily food retail market is 7.7 percent (Statistic Denmark 2016).

Parallel with this national policy for organic transition a number of municipalities have developed their own local strategies for organic conversion, and some municipalities have fostered local strategies for organic conversion. In this chapter, two of those municipalities will be explored as case studies about activities relating to organic development. The City of Copenhagen has a policy of 90 percent organic use in municipal public kitchens and the rural Lejre Municipality has declared the municipality an "organic municipality." These two cases will shed light on some of the similarities in the implementation of these programs, but they will also show differences in their approach and development planning. These case studies provide an opportunity to discuss different strategies and possible learning outcomes for implementation in other place.

Methodology

To gather information to create the cases that provide the basis for this chapter, an extensive number of interviews, observations, and visits to the Lejre and Copenhagen municipalities have been carried out. The methodological basis for the study is a qualitative insight to the network of actors and materialities important for understanding the process of converting political vision into real life action. The inspiration to work with a network- and complexity-oriented approach comes from the Science and Technology Tradition and allows the study to be complex, contradicting and hybrid, meaning it consists of both material and human actors in the networks (Latour 1999; Clarke 2005). Both places are spectacular in terms of their achievements on organic transition although very different in context, size and approach. From a research point of view they are interesting since they offer insights into the complexity of creating public attention to increase organic demand, and illustrate the possibilities for action when political and organizational will is enacted.

The study

Interviews were conducted with the mayor, the head of the Lejre Organic Municipality (LOM) project, the administrative head of the municipality, and several businesses in the municipality (farmers, food companies, and consultants). In addition, documents from Lejre and Copenhagen council meetings, newspaper articles, reports, and magazines were studied. In Copenhagen there has been continuous contact with the administration for children and with the consultancy company Copenhagen House of Food (CHF) over the last two to three years, and in 2016 a study was conducted at some of the "food schools" (Dal and Hansen 2016). The interviews were conducted primarily in 2014 (Copenhagen), 2015 (Lejre), and again in 2016 (Copenhagen and Lejre).

The case descriptions are meant to be basic introductions to the major networks supporting organics in the two municipalities. The aim of this analysis is to identify the similarities and differences in the two paths of development.

Presentation of the Cases

This chapter will discuss two case studies of the visions, strategies, and experiences focusing on organic food in two administrative and political units, Lejre and Copenhagen municipalities, with a specific on LOM and Copenhagen organic public food. In the framework of discussing common developments for change and basic differences, the chapter analyzes points of political will, network and allies, motivation and restrictions, organizational embeddedness and situated policy creation in a framework of network and complexity oriented understandings of actions (Clarke 2005).

Lejre Municipality—A Case Description

The municipality of Lejre—fifty kilometers outside greater Copenhagen—declared the town an "organic municipality" in 2012. The creation of local alliances in the rural Lejre municipality was inspired by the fact that the former minister of food reduced the support to municipalities working with organic conversion. So this change in central government inspired local governments to find ways for continuing organic transition, and it illustrates the effect of local political action toward more national-oriented activities and offers the basis of a discussion about the complex actor network involved in making changes to the food scene.

Lejre is a rural municipality close to Copenhagen, the capital of Denmark. The municipality is dominated by farming areas and small towns spread over the land without a defined city as the main urban center for activity. In 2010 the municipality had a new mayor from a left wing party (the first time the municipality has had a left wing mayor), who had an interest in organic food production and a large network of connections in sustainability, organic food, and nature

FIGURE 8.3 *The LOM logo. Image courtesy of the Municipality of Lejre.*

conservation in general. A collaboration between the mayor and representatives from two NGOs, Nature Conservation Denmark and Organic Farming Denmark, both influential advocates in Denmark for nature conservation and organic farming, agreed to create a vision they termed Lejre Organic Municipality. Figure 8.3 shows the logo developed for the initiative and illustrates different aspects of the vision and involved actors, such as the church, as an active part.

The mayor got support from the municipal council for the LOM project for a three-year period. The support from the municipal council came from a majority of the voices in the council due to an approach in line with the overall decentral and citizen oriented strategy. The city council typically consists of representatives of the political spectra from right wing to left wing parties. By communicating the project as an inclusive idea and strategy she underlined the need for political support and included the "ordinary" citizens of the municipality in the idea.

A common aim for the whole political spectrum in Lejre has been to safeguard the quality of groundwater (for drinking water) which is crucial for all parties. This has a long tradition as Lejre municipality is one of the main suppliers of drinking water for larger cities nearby including the capital Copenhagen.

A few months after the approval of the LOM visions in the municipal council the mayor created a position in the municipal government with the main aim to support and make this project happen.

In starting up the project the mayor also took action to communicate the strategy and vision internally in the administration of the municipality, and expected all heads of municipal administrations to have continuous inputs to the strategy on how their administrative area could and would contribute to the LOM ideas.

The "silo" thinking, where ideas and projects are often placed in a specific administrative unit, seemed to have a large influence on how the development and realization of political visions is brought forward (Boxelaar et al. 2006). In the case of the organic municipality the visions brought forward by the mayor were expected to permeate the whole of the municipal organization. This meant that the mayor played a quite active role in the beginning of the realization of the LOM project by, for example, demanding an immediate conversion of the municipal products served at meetings, etc., to be organic and, even more importantly, demanding a reflection of the LOM visions inside *all* administrations by organizing the agenda of regular management meetings such that LOM became a part of the discussion. This was an unusual initiative, according to the administrative director of the municipality because often in a municipal organization the mayor would not have direct influence over or interest in a specific issue to be regularly discussed at management meetings of the council.

According to the project manager of LOM these regular discussions were meeting resistance in the beginning of the LOM project because some of the administrative units (managers) did not see the relevance in discussing organic food related to their areas of influence, such as nursing homes or business development. With the continuous focus from the mayor this was kept in mind, and therefore no unit was considered free from engaging in the discussion.

The LOM Project

The head of the LOM project was, shortly after she started working in the municipality, asked to define the project more or less from scratch. In her job description there was no plan to follow or milestones to meet. It was according to her a "blank paper to fill out" although there was some overall goals. The first year concentrated on two areas: (1) on creating networks and facilitating all types of ideas that emerged from the bottom-up level—often by citizens of Lejre and (2) to involve the municipal institutions in a conversion towards more organic produce in the kitchens. The first year was therefore characterized partly by these small and diverse citizen driven initiatives such as the Lejre food community, Lejre organic garden community, the apple orchard communities, etc. In addition, the local church arranged a food festival early in the LOM project period. For the other main focus points—the public kitchens and their conversion toward organic produce—the municipality involved a consultant from Copenhagen with experience in conversion of public institutions in the capital—the Copenhagen House of Food (CHF).

The share of organic produce in public kitchens in Lejre was quite low, and according to the municipality they used only 17 percent organic products before the LOM project was initiated. After two years they increased usage to 60 percent organic products and also reduced food waste rate, which allowed them to describe the conversion as a success (LOM). The official goal of the municipality is 75 percent organic products in the public kitchens during 2016. It is not clear whether they have reached this goal yet since there has not been a final measurement (January 2017).

The Project Manager Role

The director of the LOM project began with no detailed roadmap or goal for the development and success of the LOM project. This was partly a deliberate strategy but also an expression of the new direction that needed to be created. From the conversation with the head of LOM it is clear that here was a lot of "invisible work" done in the first years of the project creating connections between people with an interest in developing common ideas, arranging communication diversity and establishing basic guidelines and funding opportunities for smaller projects by offering support for new ideas.

One quite unusual aspect of the internal organization of the LOM project was that the project manager was at the level of direct communication with the mayor. This is underlined by reflections from the administrative municipal manager in Lejre who started while the project was running in its second year. She was surprised that the LOM project manager had direct access to the mayor's office since this is quite rare in the hierarchical municipal organization. This was very obvious to her, since she also had to make room for the project even though it was not the most important or dominating agenda in the municipality.

By this example it is clear that the organizational and informal structures around an idea will have consequences on many levels in a municipal organization. It is quite obvious that if the LOM project had been "hidden away" in one of the administrative units it would probably not have had the mayor's or the head of administration's attention continuously and the life of the project might have been quite different. From the interview with the manager of the LOM project it was clear this was happening from the start since she was struggling with her organizational position during the first phase in the administration for environmental and maintenance. She felt she therefore needed to argue for her decisions and priorities with a (former) administrative manager who did not have much knowledge, vision, or mandate for the organic conversion. The only reason she was able to actually act was because of the open-door policy with the mayor's office, which the head of administration also noticed when she started in the municipality. During the next phase the LOM was moved to an administrative unit closer to the strategic decision unit and organizationally closer to the mayor.

The project manager has been playing an increasingly important role in creating the project's identity. After four years she has found a role in the municipal system where she is acting mainly as an intermediary, creating the connections between people who are recognized as resources in the LOM story. She defines her role as a facilitator to make sure the right people meet each other so they can accomplish the LOM goals. She does not see herself in a role that initiates or starts new projects but supports existing or emerging actors and projects.

Communication

One major focus along the bottom-up activities and the conversion of public kitchens has been to communicate the results to citizens inside and partners outside Lejre, which meant that even minor ideas and projects were covered in local newspapers and a magazine developed from the LOM secretariat. The Magazine Lejre is an example of the broad communication strategy that targeted all citizens to inform about the numerous ideas and initiatives related to food in the municipality.

Development—From Project to Identity

During the years of the project more focus has been put into commercial development to define Lejre as an interesting and attractive place to establish quality food companies and activities. The LOM has been created as a trademark for quality food that goes beyond the concept of organic food. A communicative strategy for more to feel included although this has not been directly stated.

The strategy seems to work in terms of external interest. Although there is a general tendency nowadays to leave the countryside and move to bigger cities, and although all municipalities struggle with their budgets and the ability to attract new companies or citizens, the LOM development seems to prove something different. For instance, the project has been developing a sense of identity for Lejre and the people in Lejre. Interviews with owners of different companies underline that they experience a change in attention and many people in Denmark know of the Lejre "Organic Municipality" identity. Place and identity are also reflected in a new pamphlet of the Lejre goals called "Our place" (LOM). Business development has been focused on creating new and relevant networks particularly inside, but also outside the municipality. One of the key tasks of the LOM project manager has been to look for and encourage interesting partnerships among the existing and new companies. This support, combined with continuous communication and the Collective Impact (CI) approach, has inspired the project leader.

Lejre: Inclusive Strategy and CI

The mayor underlined the importance of being inclusive rather than exclusive in the program strategy and this approach assured her success since she was reelected with one third of all personal voters in Lejre municipality in 2013. The inclusive strategy seems to have worked and spread to a large number of the citizens in Lejre as well.

The persons interviewed about Lejre state that the CI approach developed by Kania and Kramer (2011) was inspirational. The CI approach identifies focus points to be aware of in a highly complex management process. The major points include: (1) creating a common agenda, (2) agreeing on shared measurements, (3) support mutually reinforcing activities, (4) emphasis on continuous communication, and (5) establish backbone support (Kania and Kramer 2011).

The major focus points

In Lejre there has been a focus on securing a common agenda and supporting continuous communication as well as a support person or secretariat that can make sure the focus of a given change project is not lost. All three points in CI seems to be very similar to the process in Lejre while another point mentioned in the CI approach is the shared measurement point. Here it is a bit more unclear whether there is a shared measurement goal and what type of goal that is. There was, for instance, at one point an announcement that the percentage of organic farmland was the highest in the country (11 percent), while there was less attention to the fact that almost none of the larger farms in the municipality have converted to organic. An effective and targeted campaign ideally results in concrete conversion of conventional farms. The public kitchens conversion is another point where the measurements are not followed closely, although there has been a statement of a goal of 75 percent organic food in the public institutions in 2016; this has not yet been communicated by measurements.

The LOM Project and Its Success

Summing up the LOM and the process of development from the LOM project to creation of identity markers for the municipality, this section describes the development goals attained.

The shift from working on and within the municipality to acting much more as a network creator among private entrepreneurs has become more obvious during the three years of the project period. The focus of LOM is now more focused to enable development in food production by supporting small organic producers or new initiatives—such as a new collaboration with an esteemed restaurateur who has invested his time and money in farmland to grow his own food for his (Michelin starred) restaurants in Copenhagen, and creating a program for educating chefs in food quality and product development using locally grown food. The project has therefore undergone development from creating more public and municipal connections to the public institutions and the farmland supplying public places to a much more entrepreneurial oriented sphere of companies and restaurants seeing the potential of using Lejre's geographical and territorial placement, but also reputation, as a progressive place to establish themselves and for marketing their products as originating within the territory.

The LOM project manager is therefore also increasingly focusing on creating these new networks among entrepreneurs and (small) organic farmers rather that talking to the conventional farmers or the public institutions in the area. According to her, the public situation is not yet mature for this type of thinking in larger scale. The development in the rural municipality is quite different from the organic development in the Copenhagen municipality and the collaboration with CHF was therefore not without challenges although several institutions in Lejre have been converting their menus and product lines to be at least 50–75 percent organic.

The LOM project ended after three years but the Lejre municipal council decided to integrate the project into the more permanent strategy work of the municipality, which underlines the success of the project. Since the project period the head of LOM has been moved into a business development administration with an action area focused on sustainability and climate change and she has thus gained more formal power in the municipal unit.

In the next section we will turn to another municipal case study from the new Danish food scene, an example from the city of Copenhagen. It has proven to be a game changer for the schools of Copenhagen and has changed the school lunch history in Denmark, which for the last fifty to seventy years (Kainulainen et al. 2012) had become dominated by lunchboxes in almost every public and private school (Lukas et al. 2012). The Copenhagen case study will show that there was a similar starting point with a political vision and an important actor taking lead on the project management, as was the case in Lejre. In Copenhagen the key actor for implementation was installed in 2007—the CHF, which received support from the head mayor of the municipality. The CHF radically reformulated school food in Copenhagen (Københavns 2007).

Contrary to the LOM project, which was oriented towards initiatives coming from citizens themselves, the CHF was very active in creating an "identity" for the new program supporting organic, high quality food in the Copenhagen schools.

The approach to organic transition in Copenhagen is focused on procurement, food production and service, meals, food identity markers and the cultural aspects of meals, which will be discussed below.

Copenhagen Public Food—The Exclusive Strategy and a Strong Identity

The capital of Denmark, Copenhagen, is the largest city in the country and a municipality with approximately 600,000 inhabitants. In the city there are sixty schools (Københavns Kommune). This case study is quite unique within the Danish context but also across the world, since the municipality has succeeded in a transition of the public kitchens to use almost 90 percent organic products (either by weight or by price). This section will describe the Copenhagen school food case study and discuss the development of the program in Copenhagen.

Danish school food has been characterized by many past efforts to provide school food at different schools and in different municipalities of the country. Studies have showed that these have been predominantly temporary initiatives which ended due to lack of economic viability, often resulting in only a small stall to sell candy bars and fruits, if anything at all (Brinck et al. 2011; Hansen and Kristensen 2013). On the other hand, with conversion to organic, as stated by the most recent national political action plans and visions, public kitchens are regarded as a major organic "consumer" while private households are harder to regulate and have an impact on politically. As a result, public food in schools, kindergartens, and nursing homes are all strategically interesting when seeking to understand the full effects for conversion of farmland to organic.

The history in Copenhagen is an example of a successful organic school food development where the popularity of the school food has steadily increased and the political attention towards public food has changed radically since the early 2000s. The political vision of food in Copenhagen was—as in Lejre—carried out by a prominent political actor, in this case the head mayor of the municipality. In 2007 when the political focus on school food was reformulated, the head mayor was a Social Democrat who had an interest in quality food and organic production. With the backing of the city council she became an important driver in the establishment of a "Food House" in Copenhagen. This Food House was designed to raise quality of food in public institutions and thereby also to change the image of—and products in—the public kitchens.

A combination of political goals and good relations between the head of the CHF and the head mayor in Copenhagen created a basis for the very ambitious conversion of public kitchens to organic products, and today up to 90 percent of the food in public kitchens in Copenhagen is organic. This is a quite unique story and touches upon many aspects of the political will, the identity work of CHF and public kitchens being able and willing to change. For several of the kitchens this has been a long and hard process involving not only buying different food products but also gaining new competencies and working with completely different menu plans. The entire transition has been taking place using the same budget, meaning that the menus had to be drastically changed, for example, the content of meat and processed food products had to be reduced in order to keep expenses low. Figure 8.4

FIGURE 8.4 *The Copenhagen House of Food. Image courtesy of the Copenhagen House of Food.*

shows the Copenhagen House of Food from outside. The house is situated in the former meat packaging district which is now a lively area for popular quality restaurants and bars.

The head mayor who for many years had been involved in food-related political agendas was able to get the city council to support the initiative which created the "House of Food" in Copenhagen as a power center for organic conversion in public kitchens. The organizational structure of the Food House was unusual in a Danish municipal context since the House was connected directly to the financial department of the municipality rather than a more specialized department. As a result, the head mayor strongly influenced the decisions regarding the CHF. The CHF was assigned to create a goal for increasing organic and quality food in the public institutions. Organic food in schools had been on the agenda before, but the establishment of the CHF boosted the effort dramatically. Placed in one of the former meat processing buildings in the old meat district that at that time (2007) was becoming an area of progressive development and gentrification, the presence of the House of Food signaled high-quality, progressive development without compromising the authentic "look" of the raw factory buildings. The CHF soon became a strong identity marker and the manager of the House was noticed nationally and internationally because of her goals and uncompromising attitude.

The CHF has a number of different activities, one of the most important being courses and network building for kitchen staff in the municipality. In addition, consulting and creation of

FIGURE 8.5 *The yearly awards dinner for public kitchens at the town hall in Copenhagen. Image courtesy of the Municipality of Copenhagen.*

conferences to encourage discussion about food in a broader critical perspective with important food actors at all levels became important to the strategy. Another of the important goals of the CHF is to increase the prestige of the work done in the public kitchens. Previously, public food did not get a lot of attention and certainly no prestige. Frequently the professional kitchen staff received criticism for the food they prepared and did not receive positive acknowledgement. The House of Food has been working hard to change that view of the public kitchens by focusing on staff competence development and creating competition and yearly awards among the kitchen staff.

During the last few years the CHF has widened the strategy partly because their work in Copenhagen is now more on a maintenance level and partly because the House itself is now a private consultancy not economically and politically financed and supported with resources by the municipality. This development occurred after years of political influence by CHF decreased because the head mayor was replaced by a new mayor with no specific interest in food (the election of a new mayor from the same political party happened in 2010). The management of CHF worked hard to convince the new head mayor to continue the work with organic transition.

Since then food has been gaining steadily and increasing attention in Copenhagen and the new head mayor is warmly supporting the food related initiatives—including the public kitchens conversion. Even opposition politicians are now positive about organic food in public institutions.

This development is partly due to the CHF identity work and lobbying in the municipality. Figure 8.5 shows the yearly award dinner and illustrates the success. The town hall is full, people are dressed up, and many politicians participate in the event, which strengthens the message of acknowledgement to the kitchen staff. This was from the beginning a purpose in itself for the Copenhagen House of Food in order to motivate the kitchen staff to work with the transition.

The Project Manager Role

Central to the development of public meals in Copenhagen is the project manager, who is charismatic. She has been in front of the development driven by CHF and was in close collaboration with the head mayor at the beginning of the visionary project. She has been able to create a large network of central actors from the Danish food scene including Organic Denmark as well as newer stakeholders involved in the food scene such as labor unions and the politicians who initially were against the CHF establishment. In some respects it is a similar development to that of Lejre municipality. In both cases the central drivers for the development have had a more or less "direct line" to the mayor and thereby could secure attention for the agenda. But in contrast to Lejre, the CHF has been able to act upon a predeveloped agenda from the beginning, and with political backup allowing them to implement radical change of the public food systems. Even though there was a clear and value-driven agenda, there were several critical voices inside the municipal organization (both administrative departments and kitchens) because normal communication channels were challenged during the process of change. At the same time the well-defined values and goals also created a strong identity for the CHF and the values communicated, which has attracted a lot of attention both within and outside of Copenhagen. With her uncompromising style of management the program manager has been able to attract attention and funding to the CHF year to year, while also creating people critical of the program because they did not feel that they had a say in redefining the system.

Top-Down Process

In Copenhagen the activities in the CHF were primarily focused on the redefinition of public food already served in Copenhagen (which was also to some point organic). The strategy was to completely redefine the existing program and create a new food regime that was applied with political support and with the dedication of the head mayor. This strategy led to a contentious period which resulted in a long-term process of challenging the central kitchens responsible for the majority of the food production in the public institutions in Copenhagen. There was a strong vision and a clear strategy to make a radical change by the CHF. The slogan of the CHF, "transition in pots and heads," confirms this approach by emphasizing that only organic products are to be used and also the need for a change of attitude in order to understand the "right" way of working with food production in the public kitchens. As a result, public food transition based on this development has had a huge success due to a clear vision and a top-down approach guided by the CHF goals and supported by the politicians of the municipality.

Organic Transition of Public Institutions

The two case studies of municipal transition to organics described above have similarities and differences to each other.

In Copenhagen, the transition process for the program was initiated by the head mayor and implemented by the CHF and carried out as a top-down transition rather than a "collective impact" process of inclusion of common agendas. The overall driving force in Copenhagen was a political goal to change the system radically with multiple aims. The new agenda was defined through the establishment of the CHF and guided by a strong vision. But it also created an exclusive (organic and gastronomical) approach rather than one which involved affected stakeholders.

From the beginning the strategy was exclusive rather than inclusive even though one of the goals was to increase the prestige of the public kitchens and their staff.

In LOM the work was much more driven by a bottom-up and inclusive strategy because the mayor supported the program and invited all in Lejre to participate. The program was not always specific and clear in the details, thus creating room for a bottom-up approach where people and projects defined the program as it was implemented. The LOM project could be characterized as a large number of smaller initiatives rather than a single and large transitional revolution. The strategy has been voluntarily for farmers, so the term "Organic Municipality" might seem a bit misleading since the transition focused on farmers' actions, rather than those of the municipality. Only a very small number of hectares have been converted during the LOM program period, and at the moment 11 percent of Lejre farmland is organic. Although that is above average for Danish regions it has not been due to the LOM project but because Lejre already contains big organic dairy farms that converted in the 1990s.

These case studies present two interesting outcomes: (1) the Lejre mayor who spearheaded the program left the municipality recently (November 2016) and leaves behind a project and municipality with a program vision but without an organization strong enough to carry out the goals, and (2) the municipality has experienced a lot of attention from all over the world about the program and goals but interestingly, not so much from other Danish municipalities. The ideas and goals of the "Organic Municipality" may be too radical for many rural municipalities in Denmark, which are still very dominated by conventional agriculture. But it does seem as if towns and regions in other countries perceive a potential for organic development, as several African countries have sent visitors to Lejre to examine the initiatives.

These two cases illustrate different approaches towards making changes and also different possibilities for creating alliances between central actors who seek to change food production and consumption. Since the LOM project started, the small rural municipality has faced increasing attention for their project and is increasingly seen as an attractive environment for small and medium sized quality food companies to establish themselves in the municipality. The LOM project has in the last year become a more permanent part of the municipality strategy instead of being a project.

Today the two different municipal developments meet in a common vision for rural–urban developments. "The Food Community initiative" is a network of important actors involved in the development of distribution connections for produce from rural and farming based municipalities (Lejre and Bornholm primarily) and the capital Copenhagen—and also linked to the new "Greater

Copenhagen" initiative. This project involves both established actors such as CHF and LOM but also involves other stakeholders around the food production–consumption link. This greater initiative has strong connections back to the two cases described in this chapter and how the public food sector is perceived to be a driver for increasing conversion of farmland from conventional to organic.

References

Boxelaar, L., Paine, M., and Beilin, R. 2006. Community Engagement and Public Administration: Of Silos, Overlays and Technologies of Government. *Australian Journal of Public Administration* 65.

Brinck, N., Hansen, M. W., and Kristensen, N. H. 2011. Forty Days of Free School Meals as a Tool for Introducing Market-Based Healthy School Meal Systems in 35 Danish Schools. *Perspectives in Public Health* 131.

Clarke, A. E. 2005. *Situational Analysis—Grounded Theory after the Postmodern Turn*. Thousand Oaks, CA: Sage Publications.

Dal, J. K., and Hansen, M. W. 2016. *Undersøgelse af Madskoler i Københavns Kommune (Study of Food Schools in the City of Copenhagn)*. Denmark: Aalborg Universitet.

Danish for Copenhagen Municipality (Københavns Kommune). 2017. Befolkning og fremskrivninger. www.kk.dk/artikel/befolkning-og-fremskrivninger (accessed January 2017).

Devas, N., Delay, S. 2006. Local Democracy and the Challenges of Decentralising the State: An International Perspective. *Local Government Studies* 32: 5.

Hansen, M. W., and Kristensen, N. H. 2013. The Institutional Foodscapes as a Sensemaking Approach towards Food Schools. In *Making Sense of Consumption. Selections from the 2nd Nordic Conference on Consumer Research*, ed. L. Hansson, U. Holmberg, and H. Brembeck. University of Gothenburg, pp. 299–312.

Kainulainena, K., Benn, J., Fjellström, C., and Palojoki, P. 2012. Nordic Adolescents' School Lunch Patterns and Their Suggestions for Making Healthy Choices at School Easier. *Appetite* 59.

Kania, J. and Kramer, M. 2011. Collective Impact. *Stanford Social Innovation Review.* , Winter 2011, pp. 36–41.

Københavns Madhus. 2007. Redegørelse om skolemad i Københavns kommune. http://kbhmadhus.dk/media/351875/Redegorelse_om_skolemad_presse_24_08.pdf (accessed January 2017).

Latour, B. 1999. On Recalling ANT. *Sociological Review* 47.

Lejre Municipality. 2016. Vores sted (Our Place). www.vores-sted.lejre.dk/media/3402/vores-sted-en-lille-bog-om-lejre-kommune-light.pdf (accessed January 2017).

Lejre Organic Municipality. LOM Evaluation. http://denoekologiskekommune.lejre.dk/media/6262/resultater_rapport.pdf (accessed January 2017).

Lukas, M., Løes, A.-K., Nölting, B., and Strassner, C. 2012. iPOPY and Lunch Box Culture. http://orgprints.org/21756/1/EF2-2012_excerpt_ML_AKL_BN_CS.pdf

Ministry of Environment and Food. 2009. The Organic Eating Label. www.oekologisk-spisemaerke.dk (accessed January 2017).

Ministry of Environment and Food. 2012. Økologisk Handlingsplan (Organic Action Plan 2020. http://mfvm.dk/fileadmin/user_upload/FVM.dk/Dokumenter/Landbrug/Indsatser/Oekologi/Oekologisk_Handlingsplan_2020.pdf

Ministry of Environment and Food. 2015. Organic Actionplan. http://mfvm.dk/fileadmin/user_upload/FVM.dk/Dokumenter/Landbrug/Indsatser/Oekologi/OekologiplanDanmark.pdf (accessed January 2017).

Statistic Denmark. 2016. www.ds.dk/profil-af-den-oekologiske-forbruger.pdf

PART THREE

Organic Food Values: Sustainability and Social Movements

EDITORS' INTRODUCTION

"Vote with your fork," you are told: "Your choices demonstrate your values." For many consumers, it's an act of faith that food choices must articulate values and support the food production methods that uphold personal and cultural ethics. How land is farmed, fertilized, and sustained, how water is utilized, how animals are raised, how employees are treated and how fruit and vegetables are grown, harvested, stored, and prepared are processes subject to choices that may or may not uphold individual ideals and values. As consumers, we usually think of the value of food in relation to health, price, cultural appropriateness, culinary capacity, and taste. When we say we wish to "uphold our values" through food choice we might choose to buy a particular brand, a particular process (such as certified organic, "Cage-Free," or "Humanely Raised") or from a particular producer whom we know and trust. Rarely do we think about value beyond that point of purchase or even ponder the concept of value beyond personal choice and preference. Rarely do we think through how the very concept of "value" affects how we think about food.

The chapters in this section explore differing forms of value in food and farming. Starting with the foundation, land, which is often too expensive for many young farmers to purchase due to rising demand in both rural and peri-urban areas. In rural counties, older farmers may hold title to land they can no longer farm but is valued at more per acre than a young farmer can afford, while in peri-urban zones the value per acre spikes as population increases and developers convert farmland to subdivision. How can organizations and land trusts structure the transfer of

FIGURE III.1 *Organic garden at Spannocchia. Photo courtesy of Jacqueline A. Ricotta.*

land to provide fair value to all stakeholders? The second chapter examines the health value of organic foods. While consumers are convinced that organic foods are healthier, the evidence is for that belief is complicated and inconclusive. Some categories are clearly better (exposure to pesticides) but some are more nuanced, including those correlated with specific health outcomes. The next chapter asks how the processes of farming can support environmental sustainability and mitigate or adapt to climate change while simultaneously supporting the economic needs of farmers? Can the ethical and economic values demonstrated by a commitment to the "Three Ps of Sustainability" (People, Planet, Profit) provide a means to think through farm economics with organic and agroecological processes? And finally, a grounding in the more philosophical value of organics through an examination of the meaning of dirt-as-soil in Western thought. How have our conceptual constructions about soil and the land channeled how we value the dirt/soil under our feet, the dirt/soil that provides our food?

The value of land—as possession, producer, and philosophical entity—is complicated because the cultural construction of food and farming involves processes that affect each other and each element of the whole. As we examine our values, we begin to see that the construction of our sense of self is invisibly determined by our understanding of the land—the world, really—around us. We realize that how we think about the land is shaped by the same cultural categories, histories, and practices that determine how we think about ourselves.

9

Profile: Leigh Bush, Food Anthropologist

I begrudgingly left the Colorado Rockies to attend graduate school in the Midwest. Meanwhile I had grown up on the East Coast, closer to New York City than to a cow pasture. So I wasn't especially knowledgeable about farms and farming culture. During my second year in graduate school I was able to intern on a farm, where a colleague and I lived in a small shack right next to the cow pasture and several chicken coops. From six in the morning to six at night we helped make animal feed, tend chickens, move hoop houses, bale hay, and the like. One day, our farmer Adam told us a story about how he came to be a slaughterhouse and butcher shop owner in addition to a farmer. In the year 2000 around Easter, Adam had brought his herd of hogs in for slaughter. He received a small check for the hogs and headed to the store with a stomach for Easter dinner. After purchasing just the necessary household goods, he realized that there was not enough money left to buy an Easter ham. So, there he was, a hog farmer without money enough to buy a ham to celebrate Easter with his family. Because the price of meat was so low due to cheap (often artificially cheap) factory farming, Adam had to find a way to either make more money or sell his farm.

So far, Adam has taken on $800,000 worth of debt in an attempt to save his way of life by slaughtering, butchering, and selling his pasture-raised eggs and meat directly to restaurants and customers. He is hopeful that people's increased consciousness about where their meat comes from, and what industrial agriculture and confined animal feeding operations are doing to the health of our bodies and the earth, will help him repay his debt and maintain his farm. As my colleague and I worked on the farm day in and day out, we came to realize the expertise that would be lost if a farm like Adam's failed. From the precise number of animals that can be safely raised per acre without using hormones and antibiotics, to reading the weather patterns for harvesting and feed production, to crop rotation and understanding warning signs in chicken poop—from the day they were born, Adam and his children had gathered knowledge of the land and the animals. So, by the time I left, with calloused hands and new shoulder muscles, and with an understanding of how to raise, slaughter and butcher a number of animals, I could also see why Adam had worked so hard to save his farm and his family's way of life. Meanwhile, as other local farms have lost that struggle, the town where Adam lives has seen businesses shutter annually. As people have moved away, the classic Midwestern town square, once buzzing with small restaurants and local shops, has become half-empty. Instead, the people remaining eat at one of the chain stores

FIGURE 9.1 *Leigh Bush, food anthropologist. Photo courtesy of Leigh Bush.*

FIGURE 9.2 *Leigh Bush washing eggs. Photo courtesy of Leigh Bush.*

FIGURE 9.3 *Leigh Bush at the slaughterhouse. Photo courtesy of Leigh Bush.*

FIGURE 9.4 *Leigh Bush hugging a cow. Photo courtesy of Leigh Bush.*

along the four-lane highway in the next town over. While I was aware of the environmental and health costs of industrial farming, for the first time in my life I experienced all of the repercussions of losing medium- and small-scale farmers: individuals, families, communities, and municipalities are each affected in turn.

When I think of organizations like SILT, I think of my experience with Adam and his children. Bringing culture and community back to farming and farming back to culture and community is integral to feeling connected to our food and the people we depend upon for it. Unfortunately, the market economy doesn't value these significant needs, and without having my hands in the dirt, or under the rump of a healthy laying hen, or at the dinner table of a loving family, I wouldn't either.

As you read this chapter, take note of how profit seeking has had real repercussions on farmland and families in Iowa resulting in monocropping, lack of local ownership, and reduction in young peoples' ability and desire to farm. Recognize that an economic approach is an ideology rather than a science and that even then its logic is defied by farm subsidies, which are further indicative that industrial farming is not better or more profitable. Then, identify the immense challenges faced by small organizations that seek to repair land, health, and community as a result of corporate control and a market ethic.

Farming for Food or Farming for Profits?

Paul Durrenberger and Suzan Erem

Introduction

The man sitting in the iron seat ... could not see the land as it was, he could not smell the land as it smelled; his feet did not stamp the clods or feel the warmth and power of the earth. He sat in an iron seat and stepped on iron pedals.

He did not love the land, Steinbeck continues in his 1939 classic *Grapes of Wrath*, any more than the bank that owned it did. The driver explains, the bank that had hired that tractor driver to bulldoze the home of a poor tenant is not even a local bank. It gets its orders from the East and the orders are to make the land show a profit.

"Where does it stop?" the tenant asks, "Who can we shoot? I don't aim to starve to death before I kill the man that's starving me."

"I don't know, maybe there's nobody to shoot. Maybe the thing isn't men at all" (Steinbeck 1939: 52). The tenant disagrees.

"It's not like lightening or earthquakes. We've got a bad thing made by men, and by God, that's something we can change." The tenant sat in his doorway, and the driver thundered his engine and started off ... and back he came. The iron guard bit into the house-corner, crumbled the wall, and wrenched the little house from its foundation so that it fell sideways, crushed like a bug ... The tractor cut a straight line on, and the air and the ground vibrated with its thunder. The tenant man stared after it, his rifle in his hand. His wife was beside him, and the quiet children behind. And all of them stared after the tractor.

John Steinbeck, *Grapes of Wrath* (1939: 50–51)

That was Oklahoma in 1939. Cut to Iowa, 2016. Most farm houses have been bulldozed, their former yards planted to corn. The pace has slowed, but every time corn prices go up, houses come

FIGURE 9.5 *Organic farm, Lancaster County, PA, September 2010. Courtesy of Janet Chrzan.*

down. One pair of elderly sisters tells of having their farm evaluated recently. They were interested in placing an organic easement on the land before selling it to their cousin, knowing that such restriction on future use would reduce the cost for him while giving them peace of mind that the land would be treated well. But even without any talk of the easement, their 320 acres came in well below the average cost per acre in 2015. The ag specialist who gave them that figure said the two homes on the site, including the pristine, century-old Victorian with nine-foot oak doorways, worked against them as the next landowner would have bulldoze them.

Nothing has changed since 1939, except to have grown exponentially in the direction Steinbeck so aptly described. Farmers still have to show a profit. They cannot do that by growing anything but corn and soybeans. They cannot do that on a hundred and fifty acres because they cannot "cash flow it." They might be able to do it on a thousand acres, but not without expensive GPS-driven equipment. The banks and the people who own them are still very much in the picture. So Steinbeck's vision has come to Iowa with a vengeance. The tractor driver was right; there is nobody to shoot. It is not the people; it is the system they are in. How do you change that? And why would you want to?

One farm manager told us he had traveled to Portugal, the breadbasket of imperial Rome. He explained that once the Romans had used up all the land around the city, they started looking for other places—Portugal among them—to grow wheat. "Now it's all rocks. You couldn't grow anything there."

Iowa's current corn and bean monoculture started in the mid-1940s, after World War II. Farms have been expanding in size and declining in number ever since. They either turn a profit or they sell out. The process has only been operating for about seventy years. Run that process for a couple hundred years and eventually you hit bedrock. Iowa's farming eroded 5.1 tons of soil per acre per year as of 2005 (Iowa Learning Farms n.d.)—some is washed away by heavy rains; some is blown away by high winds. We cannot say exactly when, but we can say that someday Iowa will look like those stony fields in Portugal. What keeps this system propelled forward, so obvious but as invisible as air, is the cost, and value, of the land itself, the people who speculate on it and the financial world that profits from it.

Nonprofits and government agencies across the United States are spending millions—a fraction of what is spent on the prevailing commodity agriculture—retraining this next generation in skills we lost when the last generation of family farms was destroyed, skills like growing food without chemicals and fixing small farm machinery. Young people who grew up in sterile suburbs now want to get their hands dirty, work long hours and grow beautiful produce, tapping into the growing local foods movement across the country.

In 1984 Bruce Colman wrote:

A new agriculture is growing in the United States. Born partly of the environmental movement of the 1960s and 1970s, inspired partly by the unending crises to which the American farm is heir, harking back ... to the great conservation struggles of the 1930s ...

It involves people gardening without chemicals ... It involves developing new ways to own farmland and to preserve it against the land-hunger of sprawling cities, malls, and highways. It involves new attention to the oldest faults of agriculture: its self-destructive tendency to wash away and sterilize the soil, to encourage deserts to develop where people have been raising their food, and to drive the best and the brightest ... people off the land. (ix, x)

Thirty-five years later, consumers have caught up and the demand for organic and locally grown food is rising dramatically. With that is a correlating rise in the interest in this kind of farming, with organizations such as Worldwide Opportunities on Organic Farms, and Midwest Organic and Sustainable Education Service gaining new popularity.

Yet most of these farmers still do not own their land. Colman's reference to "new ways to own farmland and preserve it" is a nod to the birth of farmland preservation organizations—part of the broader land conservation movement launched in the 1970s. That is the loop that is being closed in Iowa today with the Sustainable Iowa Land Trust (SILT).

Many millennials watched their parents lose or almost lose everything in 2008. Some see that getting a college education is no guarantee of a good salary and watch lifelong careers being replaced by a gig economy. This generation has less than a one-in-ten chance of getting a job protected by a union contract. And they are taking notes. They are learning how to grow their own food, live on less and be self-sufficient. One extreme is turning into "preppers" planning for the end of cheap oil and inevitable attacks on their personal property by those who have not prepared for it. At the other end of the spectrum is a generation of welcoming, hard-working people who want to build new communities of cooperation and knowledge.

But few nonprofits or agencies are addressing the issue of land security and affordable access to the land. Instead, they assume, as all Americans have since Europeans settled this continent, that

- the market rate paid for land is just part of the farm's "business expenses"

- a bank will be involved, and that bank will charge interest for a thirty-year mortgage, and that interest will nearly double the price of that land over that time

- farmers with a down payment believe they "own" the farm, even though the 1980s farm crisis proved the fragility of that assumption

- those who cannot purchase land will lease degraded soil, one year at a time when it takes at least three years to begin to see the benefits of organic practices

It is time to jump off this assumption treadmill. Aspiring farmers and their need for secure, affordable land are why two dozen Iowa leaders came together in late 2014 to form the SILT, a new model that introduces economic equity to the conservation world. Restrictions on land offer tax benefits that conservationists know well, but when the next buyer must be a sustainable food farmer, and that restriction is forever, it becomes a game changer for the future of local foods.

The group has two approaches: using conservation easements to permanently restrict the use of farm land to sustainable food production (taking it off the market for everyone except natural food farmers), and owning farmland on which it offers affordable, long-term leases while allowing farmers to purchase the house, barns, and business. These tactics lead to structural change the group believes will last generations: no more mortgage interest on land trust farms, and reduced debt on farms protected by easements. The value of farmland will no longer be determined by how much a developer will pay for it, and a certain percentage of land in Iowa will be required by law to grow healthy food, with the coinciding public benefits.

There is no need to shoot a bank; this model is out to change the system. Several trends converge in the SILT: a land ethic, a food/organic initiative, a back to the land vision, and a critique of corporate capitalism—all of which combine to create an innovative approach to affordable, healthy land and food in a place largely vacant of both—the American Midwest. When agricultural economist Edward Jaenicke (2016) mapped organic activity across the United States the hotspots were in the Northeast and the West; the Midwest was empty.

Land Ethic

SILT aims to enact the principles of a land ethic Aldo Leopold articulated in his posthumously published 1949 book, *A Sand County Almanac*. The central tenet is that a thing is right when it contributes to the integrity and stability of a biotic community and wrong when it does not.

Writer and farmer Wendell Berry (2012) has elaborated on these ideas in his extensive writings.. He contrasts the "land community" of Leopold with corporate industrialism which is indifferent to its sources in the land and says, "There is in fact no distinction between the fate of the land and the fate of the people. When one is abused, the other suffers."

Stewardship for Berry means not only taking care of the land but also keeping people working on the land. As he told food writer Mark Bittman in a 2012 interview, "How are we going to get a population of people on the land that aren't telling their children, 'Honey, don't ever farm. Get out of here as quick as you can,' but 'There's a place for you here. You can farm'?"

Berry claims a scriptural foundation for his ethics as does David Rhoads of Chicago's Lutheran School of Theology who suggests that the concept of environmental stewardship comes from the story of creation in Genesis in which God gives people dominion over the fish of the sea, birds of the air, and the animals of the land (Gen. 1.1–2.4). The Hebrew word that has been translated as "dominion" means "to take responsibility for" rather than "to dominate." God created humans to be responsible for the well-being of the garden Earth. Thus stewardship of creation is fundamental to our humanity (Rhoads n.d.).

Wes Jackson and John Ikerd are other contemporary thinkers who have written on the topic from the perspective of sustainable agriculture and equity. Others base their ethics on a secular concern for resource management and human ecology. Still others base theirs on a rights-based approach derived from the idea of the commons, that all of the earth's resources belong by right to all of its people and that none should have privileged access. With the rights to use natural resources comes the responsibility for maintaining them, thus all are equally responsible.

Back to the Land

An earlier back to the land movement is best exemplified by Scott and Helen Nearing and popularized by their 1954 book *Living the Good Life: How to Live Simply and Sanely in a Troubled World*. Nearing became active in social movements after his dismissal from the economics faculty at the University of Pennsylvania in 1915 and then in the 1930s began homesteading in rural Vermont while publishing many books, articles, and pamphlets to encourage self-sufficiency (Saltmarsh 1991).

Others in a similar vein were Paul Corey, an author who moved from Iowa to rural New York where he built his own house and grew his own food and wrote an instructional book to help others to do the same, *Buy an Acre: America's Second Front* and similar instructional books. Or Louis Bromfield (1943, 1948) who lived in France for fifteen years before returning to rural Ohio during World War II where he restored several farms that had been destroyed by conventional agriculture. He was convinced that national greatness was based on land and the prosperity of the people who worked it.

These pioneers of a sustainability ethic were educated, articulate, and energetic enough to publish copiously. In 1968, after touring communes, Steward Brand published the first *Whole Earth Catalog*. The catalog provided information on books and products that promoted the values of self-sufficiency, ecology, alternative education, and holism and thus brought together these several strands in one place.

The land reform movement of the 1970s was an effort to move this work toward a more structural solution (Barnes 1975). And since then, Elliot Coleman, Frances Moore Lappe, Will Allen, Wes Jackson, Joel Salatin, Vandana Shiva, Michael Pollan, John Ikerd, and many others have not only spoken to these new farmers but also to the much broader audience of food consumers who have come to appreciate fresh, chemical-free food just as packaged and preserved foods were about to shove them off the grocery store shelves permanently. Cindy Isenhour (2011) has summarized recent writing in this area.

Critique of Corporate Capitalism

The critique of corporate capitalism comes from diverse sources from the nineteenth-century Karl Marx to the twentieth-century Scott Nearing and Paul Corey to the twenty-first-century Woody Tasch.

Opposed to the land ethic is the market ethic that emphasizes short-term maximization of profit for corporate owners, management, and shareholders. This is the ethic that drives agribusinesses and excludes all ecological values. Thus, the costs associated with environmental pollution can be considered external to the firm, "externalities," and not counted in the corporate budget (Bakan 2004).

Defenders of this point of view often base their rhetoric in a sense of reality that derives from economics, which anthropologist Dimitra Doukas (2003) has shown is much less a science and much more an ideology that was self-consciously developed as a justification for the rapacity of nineteenth-century precursors of today's corporations to make their actions seem natural, inevitable, and acceptable.

Related is the notion that business is the appropriate means for managing any resource. Thus America's Morrill Land Grant colleges promulgated an ideology of farmers as businessmen and the accompanying idea of farming as extraction of maximize profits by increasing productivity. Berry illustrates this further when he describes these agricultural schools' shift from the relationship a farmer may have with a particular piece of land to a science that anyone can replicate on any piece of land. Thus the Land Grant agenda fits closely with that of the burgeoning post-war agribusiness industries that produced fertilizers, herbicides, insecticides, and genetically modified organisms and provided a means for their collaboration in the development of agricultural technologies (Durrenberger and Erem 2015).

Iowa's Current Situation

Iowa is one of the centers of industrial corn and soybean production, consistently ranked no. 1 and 2 in those crops nationally (USDA 2015; Cook 2016; Iowa Corn Growers Association 2016). Nearly 60 percent of corn grown in Iowa is sold for fuel, according to the Iowa Corn Growers Association, with 21 percent going to animal feed, the second-largest market. The rest goes to the oils and sweeteners for highly processed foods with long shelf-lives implicated in America's obesity epidemic.

The average Iowa farmer is fifty-eight years old. One-half of Iowa land is owned by people over sixty-five, one-third by people over seventy-five. Land has been held and accumulated for decades, since many of these large family farms are the victors from the 1980s farm crisis.

Many older farmers consider the profit they will make cashing out their land as their retirement security, and if anything is left, their children's inheritance. Many of their children see it that way as well. Because most see the land as a commodity, not a food producing or environmental resource, even siblings can become estranged over land inheritance. If parents keep the farm in the family, issues arise once they are both dead. When dealing with multiple beneficiaries, usually siblings, it is nearly impossible not to find one who would prefer to farm the land while others insist on selling

FIGURE 9.6 *Hoophouse interior. Courtesy of Suzan Erem.*

it. This puts the farming sibling in the unenviable position of trying to buy out his or her siblings at current market value. These scenarios only get more complicated when three generations are involved.

Fully 81 percent of Iowa's farmland is rented and a large percentage of landowners live out of state. This is caused by land passing into the hands of people who do not wish to farm it, but who use it for income. Farm management companies, which earn a commission on the rent, aggregate such lands to rent one year at a time to industrial producers who then farm as much as ten thousand acres or more in a single operation. If industrial farmers have little interest in conservation, such renters have even less. In fact, the entire system from the absentee landowner to the farmer-renter is built to maximize short-term profit with little heed paid to conservation practices that might improve soil or water quality over the long term. This is not to say there are not conventional corn and bean farmers who believe in conservation, only that the system is not built to encourage it.

Pressure on Iowa's land comes from two sources: urban and rural. By 2015, people seeking agricultural land, either for income for use, had bid up the price of Iowa's best farmland beyond $10,000 per acre (and in some areas, nearly $20,000). In addition, expanding urban areas have increased the price of land for development around towns and cities.

The size of Iowa's farm operations continues to increase as the number of farms decreases. With the industrialization of agriculture, local agriculturally based economic systems have fallen apart and the small towns based on them have disappeared. Farms have become suppliers of raw materials to global corporations whose economic relationships are internal to the corporation rather than to the place. This becomes evident at the consumer end of the system by the decades-long process of unique family-owned grocery stores and restaurants being replaced by much more "efficient" and standardized chains from Applebee's to Starbucks, from Kroger's to Wal-Mart.

The intensive farming of larger and larger areas for industrial extraction has resulted in the pollution of Iowa's waters to the extent that the state's largest water works utility is suing upriver farmers for its additional expenses treating drinking water. Not uncommon are boil alerts, public swimming alerts and warnings for pregnant women and children not to drink the tap water. Taken together, these trends are deleterious to the natural environment, to the economy and to the people of the state.

At the same time there are many beginning farmers wishing to tap into a new local food market that rewards farmers who make the extra effort to produce organic, naturally grown food. The SILT was founded to bridge the gap between new sustainable food farmers who need access to land and the current system that offers them none.

SILT

Iowa is the most ecologically altered state of the United States with only one-tenth of 1 percent of the estimated 90 percent of its original prairie landscape remaining. This has spawned an environmental movement to conserve and regenerate prairie lands. Organizations such as the Iowa Native Plant Society, the Sierra Club, and Audubon work in concert with natural land trusts such as the Iowa Natural Heritage Foundation, the Nature Conservancy and Bur Oak Land Trust to preserve natural lands from development.

FIGURE 9.7 *Orchard. Courtesy of Suzan Erem.*

SILT, in contrast, is a farmland trust that conserves land *from* development but also *for* the purpose of producing healthy food for human consumption as sustainably as possible. There are a number of farmland trusts in the United States. Twenty-eight states offer public support for the purchase of agricultural easements. The Land Trust Alliance, home to almost all land trusts in the country, provides services such as accreditation and insurance against costly lawsuits. And the American Farmland Trust is a national umbrella group that provides advocacy, research and networking for farmland trusts. Land trusts use one or both means of conservation: easements and land ownership.

Conservation easements are legal documents attached to the deed that allow certain rights in the land, such as the right to develop it, to be legally isolated from others. In principle, a conservation easement endures forever, or in legal terms, in perpetuity. Landowners continue to own the property and have the ability to sell it but the next landowner must abide by the restrictions of the easement. Land trusts may also own land donated to them and then may determine the terms of the land's use within the scope of the land trust's mission. Perhaps because most natural lands do not generate income the way farmland does, natural land trusts often place easements on parcels donated to them and then sell that land, retaining enough funds, if not more, to monitor and enforce the easement into the future.

On land SILT owns, it will offer long-term affordable leases to eligible farmers giving them the security of land tenure while removing the cost of mortgage interest, which can be significant. Farmers will own the house, barn and business based on a model promoted by the national group Equity Trust. This takes that land permanently off the market and therefore out of the land speculation game while providing a modest revenue stream to cover the organization's farm management costs. On land that SILT does not own but protects through easements, the restrictions for healthy food production remove any market pressures from development or from commodity crop farming, reducing the value of the land significantly for the foreseeable future. That reduction in value makes it more affordable than any of the surrounding land and allows that farm to coexist with residential developments that may well be home to its customers.

Thus, through SILT, beginning and disadvantaged farmers may gain access to farmland without incurring the debt entailed by purchasing land at market rate. In return, they agree to dedicate themselves to the more labor-intensive farming of healthy food that benefits the general public and diversifies the local environment and economy.

Iowa averages fifty-four people per square mile, but its population is concentrated in a half-dozen metro areas spread out across a state larger than Pennsylvania or New York. The land surrounding these metro areas are the sweet spots for SILT farms but also the most at risk. The group is building relationships with developers to explore incorporating this kind of neighborhood-friendly, nature-friendly agriculture into their development plans. When developers across the country attempted this model in the 1980s and 1990s it failed largely due to the fact that homeowners' associations were in charge of managing the farms. The lack of farm management expertise among suburban homeowners contributed to the demise of this model. In the SILT model, the land allocated for a farm will be donated to SILT, whose mission and expertise will keep qualified farmers working the now-affordable land, while removing any potential for future development of that land.

Putting It All Together

Monsanto tells its farmer customers, "We feed the world," but Monsanto does not even feed Iowa, a state that imports more than 90 percent of its food (Pirog et al. 2001). Researcher Angie Tagtow (2014) has calculated that if just 123 acres of each of Iowa's ninety-nine counties were devoted to food production, every man, woman, and child in Iowa could be fed the USDA recommendation of fruits and vegetables. A similar calculus for the whole planet would suggest that the best way to feed the world is to foster the consumption of locally produced foods. Such a policy, while ethical on many dimensions, is counter to US foreign and domestic policies such as the US Farm Bill passed every five years that encourages the large-scale production of commodity crops (corn, wheat, rice, peanuts, soybeans) over local food. These policies are promoted by both the agri-industry corporations and the Morrill Land Grant institutions. The problem of world hunger requires not greater emphasis on industrial production of crops but rather people having access to land and water so that they can produce food for their own locales. The market did not determine the current system so much as government policy including massive subsidies and lucrative crop insurance for commodities provided for in the federal farm bill. But government policy can also make it easier for landowners to protect their land for the public good, in this case healthy, ripe food close at hand for schools, hospitals, nursing homes, and grocery stores. A generous federal tax deduction for land conservation did just that in late 2015.

Mark Bittman (2013) suggests that the industrial food production system uses 70 percent of agricultural resources to produce 30 percent of the world's food while small household-level producers use 30 percent of the resources to produce 70 percent of the food and radically reduce transportation costs by producing locally. If productivity is calculated as people fed from an area of land rather than units of crops produced, then the United States ranks behind China, India, and the world average.

But he also points out a great power imbalance between small producers and agri-industry and their allies. Thus, one function of SILT and other such land trusts is to address that imbalance in order to improve the world's system of food production so that it can better feed people rather than produce money for the few. Unfortunately, addressing this imbalance means reconstructing much of the local food infrastructure that has been dismantled due to vertical integration of industrial food systems. That reconstruction is happening across the country with Community Supported Agriculture, food and producer co-ops and food hubs, but it is a mighty job.

Given SILT's roots in the land ethic, organics, and the critique of corporate capitalism, this sounds pretty utopian. Yet in less than two years, SILT has entered into conversations with more than fifty landowners for more than 5,000 acres of land to permanently protect for healthy food production. These Iowans are able to see the world through an ethic of meeting human needs rather than maximizing profit. When we see land for its value of growing food instead of growing money, the problems and solutions get radically redefined.

And Iowans are not the only ones. SILT is modeled in part on Puget Consumers Co-op Land Trust in Seattle and with much help from the highly esteemed Vermont Land Trust. And since its inception, SILT has received requests from Minnesota, Pennsylvania, Missouri and South Dakota asking about its model or if it can protect farmland in those states. There is a hunger for a new way

FIGURE 9.8 *Farm, Chester County, PA, September 2010. Courtesy of Janet Chrzan.*

of looking at our land, and if models such as these are not available, landowners will have only two choices. As one SILT landowner put it, they will only be able to "sell or rent to the highest bidder, or leave their land to their non-farming children to do the same."

SILT Case Study

There is not a single farm that can tell the whole story, so we will offer here an aggregate of the situation facing young food farmers and aging landowners and how SILT is helping solve the problem.

Jason lives in Des Moines and is getting his business degree at a community college. During summers in high school Jason volunteered on local vegetable farms and thinks he may want to go into farming. At the farmers' market he sees people paying 50 percent more for certified organic vegetables, or buying ground cherries or paw paws and other things he has never heard of for more than they would pay for blueberries or strawberries at the store. He reads that high-end restaurants in town are promoting local farms and charging big bucks for tiny plates of fancy greens. Acorn-fed pork and grass-fed beef are fetching twice the price as the kind grown in the confinement buildings he sees in the countryside.

When Jason decides to find some land to farm, he cannot. The small parcels close to Des Moines are expensive, driven up by developers who plan to build suburban homes there someday. Farther out, land sells in parcels of hundreds of acres, so even though it costs less per acre, it is still more than anything he could afford or need. He does not know farmers willing to rent or eventually sell to him. His family does not own farmland. He is stuck.

Meanwhile, forty-five miles away Cheryl is looking over her 200 acres of conventional corn and bean wondering what will happen when she dies. She has no kids, and her nephews seem bent on someday selling the farm her grandfather settled.

She attends a SILT Showcase Day to hear an attorney and an appraiser and meet other landowners in the same fix. She invites a SILT representative to walk her land. She listens intently to the possibility of rotational hog, sheep, or goat grazing in her oak savanna, using livestock to fertilize pastures that would then become vegetable fields, planting orchards on slopes. She keeps

FIGURE 9.9 *Apples. Courtesy of Suzan Erem.*

asking, "Once I put an easement on this no one can spray chemicals again? They can only grow *food*?" The SILT representative explains that the land trust visits the property every year, talks with the farmer, and makes sure the land is used for healthy food production. If and when the owner sells that land, SILT will work with the new owners to be sure they follow the same rules.

Cheryl gets it. When she places this easement on this land, no one will ever be able to develop it or farm it conventionally again. The value—and the cost—of that land may have just dropped in half or more for the next buyer. Cheryl can tap up to twenty years of tax benefits to help offset the loss in value of the land, but she will also enjoy peace of mind that the land will be protected for years to come. Plus, her nieces and nephews probably will not be interested anymore so she will be free to sell it to someone who cares about it the same way she does.

Jason contacts SILT looking for land and SILT connects him with Cheryl. They hit it off. Now Cheryl has a young farmer who can begin learning to grow food and Jason has a low-risk opportunity to try farming on a few acres. Cheryl has a young person who wants to learn about her farm and Jason has the hope of buying her place someday at an affordable price. In a few years, maybe Jason will be grazing animals, growing fruit and vegetables, and buying the property on contract at half the price of land around it. Cheryl will have a decent income from the agreement and a nicer view out her kitchen window.

Affordable land for people willing to grow food for their communities. Values that factor in environmental and social health. Particular circumstances that dictate particular needs, not a cookie cutter approach. These are things that make no sense to economists or the policy makers who listen to them. They are just common sense to regular people. Unfortunately, common sense, as the saying goes, is not so common. But, as Steinbeck pointed out in his novel, you cannot just shoot the system or the people in it. To make the results different, you have to change the system, a system it took hundreds of years to create. SILT is one example of taking the first steps on the long road to doing just that.

References

Adams, J. 2003. Introduction. In *Fighting for the Farm: Rural America Transformed*, ed. Jane Adams, 1–24. Philadelphia: University of Pennsylvania Press.

Bakan, J. 2004. *The Corporation: The Pathological Pursuit of Profit and Power*. New York: Free Press.

Barnes, P., ed. 1975. *The People's Land. A Reader on Land Reform in the United States*. Pennsylvania: Rodale Press Book Division.

Berry, W. 2012. It All Turns on Affection. *National Endowment for the Humanities Jefferson Lecture*, 9–40. Berkeley: Counterpoint.

Bittman, M. 2012. The Thing (or Things) about Wendell Berry. *Humanities* 33. www.neh.gov/humanities/2012/mayjune/feature/the-thing-or-things-about-wendell-berry

Bittman, M. 2013. How To Feed the World. *New York Times* Opinion. October 14.

Broomfield, L. 1943. *Pleasant Valley*. New York: Harper.

Broomfield, L. 1948. *Out of the Earth*. New York: Harper.

Colman, B. 1984. Preface and Acknowledgements. In *Meeting the Expectations of the Land: Essays in Sustainable Agriculture and Stewardship*, ed. Wes Jackson, Wendell Berry, and Bruce Colman, ix–xi. San Francisco: North Point Press.

Cook, R. 2016. States That Produce the Most Corn. *Beef2Live*. http://beef2live.com/story-states-produce-corn-0-107129 (accessed September 30, 2016).

Corey, P. 1944. *Buy an Acre: America's Second Front*. New York: Dial Press.

Doukas, D. 2003. *Worked Over: The Corporate Sabotage of an American Community*. Ithaca: Cornell University Press.

Durrenberger, E. P., and Erem, S. 2015. *Anthropology Unbound: A Field-Guide to the 21st Century*. New York: Oxford University Press.

Iowa Corn Growers Association. 2016. Corn Facts. www.iowacorn.org/media-page/corn-facts/ (accessed September 30, 2016).

Iowa Learning Farms. The Cost of Soil Erosion. www.iowalearningfarms.org/files/page/files/Cost_of_Eroded_Soil.pdf (accessed September 27, 2016).

Isenhour, C. 2011. Can Consumer Demand Deliver Sustainable Food? Environment and Society. Advances in Research 2: 5–28.

Jaenicke, E. 2016. U.S. Organic Hotspots and Their Benefit to Local Economies. A White Paper prepared for the Organic Trade Association. https://ota.com/sites/default/files/indexed_files/OTA-HotSpotsWhitePaper-OnlineVersion.pdf (accessed June 7, 2016).

Lave, J. 1988. *Cognition in Practice: Mind, Mathematics and Culture in Everyday Life*. New York: Cambridge University Press.

Nearing, S., and Nearing, H. 1970. *Living the Good Live: How to Live Sanely and Simply in a Troubled World*. New York: Schocken.

Nestle, M. 2007. *Food Politics: How the Food Industry Influences Nutrition and Health*, 2nd edn. Berkeley: University of California Press.

Newman, K. 1999. *Falling from Grace: Downward Mobility in the Age of Affluence*. Berkeley: University of California Press.

Pirog, R. S., Timothy, V. P., Enshayan, K., and Cook, E. 2001. Food, Fuel, and Freeways: An Iowa Perspective on How Far Food Travels, Fuel Usage, and Greenhouse Gas Emissions. *Leopold Center Publications and Papers*, Paper 3. http://lib.dr.iastate.edu/leopold_pubspapers/3 (accessed September 30, 2016).

Rhoads, D. Stewardship of Creation. www.lutheransrestoringcreation.org/the-stewardship-of-creation (accessed September 30, 2016).

Saltmarsh, J. A. 1991. *Scott Nearing: The Making of a Homesteader*. White River Junction, VT: Chelsea Green.

Schlosser, E. 2002. *Fast Food Nation: The Dark Side of the All-American Meal*. New York: Harper.

Steinbeck, J. 1939. *The Grapes of Wrath*. NY. MacMillan

Tagtow, A. 2014. *Healthy Food, Healthy Iowans, Healthy Communities. Part 1. Community Food Systems: A Primer for Local Public Health Agencies*. Des Moines, IA: Iowa Department of Public Health. www.idph.state.ia.us/IDPHChannelsService/file.ashx?file=CF4938B9-CD57-4404-86B1-81AA38E54FA6 (accessed June 22, 2016).

Tasch, W. 2008. *Inquiries into the Nature of Slow Money: Investing as if Food, Farms, and Fertility Mattered*. White River Junction, VT: Chelsea Green.

Thu, K. M., and Durrenberger, E. P. 1998. Introduction. In *Pigs, Profits and Rural Communities*, ed. Kendall M. Thu and E. Paul Durrenberger, 1–20. Albany: State University of New York Press.

US Department of Agriculture. 2015. Iowa's Rank in United States Agriculture. www.nass.usda.gov/Statistics_by_State/Iowa/Publications/Rankings/2015%20Rankings.pdf (accessed September 30, 2016).

Weiss, B. 2016. *Real Pigs: Shifting Values in the Field of Local Pork*. Brad Weiss, Durhum: Duke University Press.

10

Profile: Erika Tapp, Concerned Consumer

Like many others, I long assumed that "organic" means healthy. As Gadhoke and Brenton discuss in this chapter, gauging the true health benefits of eating organic can be hard to identify based on published studies. As a consumer, I have become lost in the sea of definitions of organic, the dubious claims of the "Dirty Dozen" adherents, and, of course, having a finite amount of money to spend on food of any variety. I want to eat healthy food that does not hurt the environment or farm workers. But what exactly does that mean?

At the end of the day, I am far less concerned with the health benefits of eating organic, in part because saying something is organic doesn't really mean anything to me. I live in the modern world with questionable water quality, air pollution, and dangerous chemicals used in beauty products. I also drink alcohol and enjoy pizza a little too much. For me, those types of issues affect my health far more than pesticides from produce (a significant portion of which can be washed away with proper cleaning). It is clear to me that large-scale conventional agriculture takes a huge toll on our land and water. This type of supply chain also almost invariably means that produce travels great distances. The carbon footprint from conventional agriculture is huge. Within my value system, the environmental impact of big agriculture concerns me far more than any particular health concerns around pesticide use per se.

There does seem to be some universal acceptance that produce is more nutritious the more recently it was harvested. So, I first focus on buying locally, then organically if that is affordable.

For me, meat is a completely different story than produce. Having read about the commercial meat industry and its brutal treatment of animals, excessive use of antibiotics, and environmental havoc it causes, I decided that I would only purchase antibiotic-free meat and dairy, with a preference for local and free-range. But I don't believe any of those three categories specifically signify "organic." With this change, I have noticed health benefits. After years of chronic sinus infections, they disappeared. And when I do need antibiotics, I take a shorter course or a less powerful type. I am not a scientist, so can't possibly tell you if the two are correlated. That said, I feel the benefits of antibiotic-free meat and dairy are well worth the extra expense or trouble.

FIGURE 10.1 *Erika choosing pickles at the farmers' market. Photo courtesy of Janet Chrzan.*

FIGURE 10.2 *Pole beans from the farmers' market. Photo courtesy of Janet Chrzan.*

FIGURE 10.3 *Radishes at the farmers' market. Photo courtesy of Janet Chrzan.*

Health Consequences and Perceptions of Organic Foods: A Synthesis of the Scientific Evidence

Preety Gadhoke and Barrett P. Brenton

Introduction

Our approach to studying the health value of organic foods is based both in the political-economic and sociocultural constructions of foodways and their embodied burdens on population well-being. As medical and nutritional anthropologists and public health professionals working with displaced and indigenous populations in the United States and globally (Gadhoke and Brenton 2015, 2016, 2017), we observe how organic foods mediate the pathway between one's social status and vitality. Along this pathway lie barriers and facilitators to achieving both social status and vitality, such as poverty, education, unemployment, housing, and access to and availability of nutritious foods. In our ethnographic fieldwork in the United States, specifically, some women have expressed organic foods as a status symbol, and opportunity for healthy eating, yet recognize the socioeconomic barriers and limitations between the haves and have nots. Our work has also shown that children are perceptive of the challenges and advantages of healthy eating (Gadhoke and Brenton 2015).

We are not saying that organic foods are not available, but we are stating that accessibility is not equal because of shear economies of scale[1] wherein the current overall cost of affordable organic food limits that access. We advocate for the precautionary principle on existing evidence-based public health practice on organic foods, and for structures that support a more holistic understanding and systematically defining its use and benefits. Overall it cannot be underscored enough that the public must be engaged as key stakeholders central to this process.

From the other chapters in this volume, the rising multisector interest among stakeholders is apparent as it relates to the overall significance of organic foods in human diets (OTC 2016). This attention has been broadly linked to the value of food defined as "organic" compared to those which are "conventional" or nonorganic. The focus of this chapter is on assessing the value of organic foods in terms of health risks and benefits pertaining to nutritional composition and food safety of organic foods, and their relationship to sustainable food systems and the environment. To accomplish this it is organized into the following subsections: "Evidence-Based Science on Organic Foods"; "Food Quality and Safety"; "Nutritional Value and Health Benefits"; "Consumer Perceptions and Outcomes"; and "Organic Foods and Healthy Futures."

Evidence-Based Science on Organic Foods

In order to assess the value of organic foods, it is important to critically evaluate the preponderance of scientific evidence. For that, we go to published and unpublished (or gray)[2] literature. To be published, studies can be peer-reviewed or not, and this disparity is referred to as publication bias. Unpublished or gray literature typically refers to institutional products like white reports or newsletters on case studies that are not peer-reviewed. However, excluding all gray literature from our conversation about organic foods that still meets scientific research standards would be missing a large body of work that has shown evidence for the value of organic foods, benefits and risks.

Within the world of published literature, meta-analysis[3] is considered the most rigorous form of synthesizing current knowledge on a topic. We scrutinized five large *meta-analyses* of literally thousands of articles, altogether, on organic foods that report inconclusively the scientific evidence of enhanced food quality, food safety, or health advantage in favor of organic over conventional foods (Dangour et al. 2009a; Huber et al. 2011; Smith-Spangler et al. 2012; Reganold and Wachter 2016; Brantsæter et al. 2017). That said, some studies do report higher levels of nutritionally desirable compounds, such as vitamins, micronutrients, and minerals, in organic foods through low input farming systems, such as those supported and adopted by the EU. In turn, this is contributing to favorable rising demands and price premiums for organic foods, which have also been closely associated with positive consumer perceptions of organic foods as nutritious and healthy (Leifert et al. 2007).

There is also evidence for reduction in exposure to pesticides, and less contact with antibiotic-resistant bacteria in pork and poultry. To lower antibiotic resistance risks the EU has sharply restricted the use of antibiotics in animal feed (Go Green, 2015). Therefore, the overall health benefits of organic food are potentially great (Huber et al. 2011; Smith-Spangler et al. 2012; Brantsæter et al. 2017). Regular consumers of organic foods have been shown to be wealthier and have healthier lifestyles (Strassner et al. 2015). This complicates any comparative studies on the health benefits of organic foods since a "healthier" and more well-nourished population may be seeing benefits and outcomes from other associated behaviors (e.g., higher socio-economic status, higher activity levels, greater access to resources, healthier food choices, higher health status).

That being said, a real challenge to increase the economies of scale of organic farming is political will within our governance structures to not only support organic farming practices but also fund research to better understand its costs and benefits. Due to the limited nature of studies on the topic, there is a need for more longitudinal and controlled studies that are designed to accurately make such comparative assessments reliable. Without such evidence-based scientific evaluations consumers are left making healthy food choices informed by anecdotal "post-truth" narratives that tout the benefits of one over the other. Organic and conventional food systems both benefit from increasing or maintaining consumer confidence in their products. This ideological divide of food beliefs has led to active debate over this topic in peer-reviewed scientific publications, gray literature reports, and unfiltered social media accounts. As will be noted below, is not to say that there are no advantages to organic food production and diets in the long term, and as yet there are no established negative consequences associated with organic food consumption either. However, we must take into account that evidence showing a comparative nutritional or health-promoting advantage of organic foods may appear to be negligible in well-nourished healthy populations utilizing standardized dietary assessments. This is because although nutrient levels may be recorded as being adequate from food intake data, it does not differentiate the quality or origin of those food items. In fact, this process may even mask the potential evidence for the long-term benefits of organic food in the diet.

Food Quality and Safety

Food quality incorporates a number of factors related to foods suitability in the diet, ranging from its acceptance based on sensory organoleptic characteristics (e.g., taste, texture, appearance) to perceived health value and nutritional composition. Sensory dimensions have been one of many areas of research in comparing organic to conventional food (Gibbon 2009). Similarly, food safety integrates perceptions and risks related to food production and handling from farm to table that may raise concerns about food borne illness (e.g., salmonella, *E. coli*), environmental contamination (e.g., pesticide residues), and the use of biotechnology (e.g., GMOs). Several risks are lowered through methods of organic food production including fewer agrochemical residues, heavy metals, radioactive nuclides, and lower levels of nitrates, which are typically added to conventional animal feeds and veterinary drugs. Yet, it should be noted that while "organic" production may be safer, that does not equate with "safe," given the high variability in organic food production practices. Studies argue for cleaner agricultural production practices to "minimize" food contaminants in both organic and conventional practices (Magkos et al. 2003).

To address this complex food systems web one can begin by anchoring comparative studies in both United States and international evidence-based "neutral" categories for nutrition and food safety standards. A clear picture of these standards is often difficult to obtain as one negotiates multiple government agencies with often inconsistent policies. In the United States, for example, one must traverse three federal agencies to understand how food safety risk is assessed. For example, for pesticide residues the Environmental Protection Agency (EPA) determines tolerance levels, the Food and Drug Agency (FDA) monitors and enforces those levels for exposure to foods, except for meat, poultry and some egg products which are monitored by the US Department of Agriculture (USDA). Therefore, while the USDA oversees the National Organic Program, which regulates the labeling and certification of organic products, it still requires some interagency cooperation with the EPA and FDA to establish and monitor the list of allowed and prohibited items in the production and processing of organic foods.

A joint FAO/WHO Food Standards Programme implements the Codex Alimentarius Commission. Its purpose is to protect the health of consumers and to ensure fair practices in the food trade. The *Codex Alimentarius*, or "Food Code," is a collection of internationally adopted food standards related to food quality, safety and production. It provides, for example, international guidance for establishing advisory provisions for labeling organic foods (FAO/WHO 2001).

Despite the myriad definitions for organic foods from legal to social, as seen throughout this volume, no certification standards explicitly state that organic foods are healthier, more nutritious, or safer than conventional foods. It is, however, generally understood that they are grown more sustainably and safer through the use of approved agricultural methods that support ecological balance and conserve biodiversity. In the United States, for a product to be labeled USDA Organic–certified, as defined and mandated by federal laws, 95 percent or more of its ingredients must be free of a long list of synthetic chemicals (petrochemical fertilizers, pesticides, etc.), and not be genetically altered through biotechnology (no GMOs), but are still considered safe for use in conventional agricultural production and processing. So therein lies a conundrum in comparing the two systems, and confusion among consumers, wherein one type of food is assumed to be as safe and nutritious as the other.

Industrial agricultural production of organic foods, much like in its conventional counterpart, has raised sustainability concerns. While comparisons may show little difference at this time, the advantage of the organic system through its use of fewer synthetic agrochemicals, resource cycling of water and soils, and conservation of biodiversity may be more beneficial and sustainable in the long run (Delate et al. 2013). Some evidence does support the theory that organic production on small farms can increase yields, improve soil fertility, and enhance water-holding capacity (Strassner et al. 2015), thereby reducing negative externalities[4] from using (conventional) synthetic compounds (Hall et al. 1989). It is important to note that in resource-poor settings such a system can also increase food security and food safety (Hall et al. 1989). All in all, USDA Organic–labeled foods, for example, should have been produced and processed in a way that integrates methods and practices that at least foster and support sustainable agriculture. Therefore, it is not surprising that the consumption patterns of organic consumers appear to be very similar to the sustainable diet concept promoted by the FAO (Burlingame 2012; Strassner et al. 2015). This links to our discussion below on consumer preference. The concept of an organic food value chain can thus be linked to broader beliefs about sustainable food production, consumption, and safety.

Nutritional Value and Health Benefits

In the United States, dietary guidelines are established cooperatively with the US Department of Health and Human Services (DHHS) and USDA. The 2015–2020 Dietary Guidelines for Americans includes absolutely no mention of the word "organic" (DHHS and USDA 2015). Whether this is because of the lobbying interest of conventional foods' agribusiness constituents, or the lack of current evidence showing any dietary benefit to organic food consumption, is unclear.

Despite inconclusive evidence highlighted in this chapter, executive branch agencies, such as the DHHS, have taken an authoritative step by utilizing a conservative approach towards promoting organic food production. To understand the public health perspective on organic food, the reader does not need to go any further than the Centers for Disease Control and Prevention (CDC).[5] On the topic of organic foods, the CDC advocates for "a food system that provides healthy, sustainable choices, minimizes environmental impacts, and serves as a model for the broader public health community. It is up to you, as a purchaser and consumer, to consider the impact of food from seed to table. Choosing local, healthy, environmentally responsible food helps promote personal health as well as the overall health of the community" (CDC 2018). The CDC calls for the choice of food that:

- Does not harm the environment
- Supports and preserves rural communities
- Is healthy and nutritious to eat
- Respects farm animals
- Provides farmers with a fair wage
- Is free of added toxins
- Is grown in the local community
- Does not harm the health of farm workers

From the CDC promotional language, it is evident that the responsibility of sending health messages to the public includes a conservative or precautionary message.

Internationally, we can look to the UN for dietary guidelines established by both the WHO and FAO. They are meant to be food-based and modified for specific countries or regions. A review of various national dietary guidelines by Fischer and Garnett (2016) demonstrated a wide variety of interpretations for what constitutes a healthy diet. German, Dutch, and British guidelines did mention organically produced foods, but it was primarily in the context of supporting sustainable environments, not because of any inherent health benefits of organic food itself.

A summary report by Reganold and Wachter (2016) reviewed evidence for organic food being more nutritious, particularly with higher micronutrient concentrations of vitamin C, total antioxidants and total omega-3 fatty acids, and higher omega-3 to -6 ratios. Worthington (2001) also reports studies that have shown increased levels of the micronutrients iron, magnesium, phosphorus, and vitamin C, as well as less nitrates in organic crops as compared to conventional crops. In addition, a meta-analysis of milk has shown that organic milk has a more desirable fatty acid composition, and significantly higher α-tocopherol and iron (Fe), but lower iodine (I) and selenium (Se) concentrations (Średnicka-Tober et al. 2016).

These results run counter to Dangour et al. (2009a) who concluded from a systematic review of studies that there was no evidence for a difference in nutrient quality between organically and conventionally produced foods. They argued that the distinctions that were noted mostly resulted from differences in production methods. It was noted, however, that organic foods were not inferior to conventional ones in terms of nutrient content. A response by Benbrook et al. (2009) noted methodological flaws in Dangour et al.'s (2009a) selection process that removed work where the organic certifying body was unclear or that was considered gray or unpublished literature. Thus, a number of evidence-based data examples were removed from the study that clearly supported research that some nutrients were higher in organic foods. They cited in particular a report produced by the Organic Center (Benbrook et al. 2008) which showed that all varieties of organic wheat contain nutritionally significant amounts of phenolic acids, flavonoids, and carotenoids, including lutein, zeaxanthin, and β-carotene when compared to conventionally grown counterparts.

An additional response to Dangour et al. (2009a) by Gibbon (2009) commented that nutrient content was not the primary issue in consumer preference, although this was not really the point of the systematic review. These critiques both drew a response by Dangour et al. (2009b) who simply stated, "Greater interaction between agricultural scientists and public health nutrition researchers is needed to improve the quality of the existing evidence base and to understand better the strengths and limitations of, and the conclusions that can be validly drawn from, established scientific methods" (p. 1701). It has also been noted that a great deal of the nutrient variation may be explained by differences in plant variety, soil and crop fertilization, ripening stage, plant age at harvest, and weather conditions (Huber et al. 2011).

A recent systematic review by Smith-Spangler and colleagues (2012) also argued that the published literature lacks a strong consensus or demonstrated evidence that organic foods are any more nutritious than nonorganic or conventional foods. However, they do suggest that organic foods may reduce exposure to pesticide residues and antibiotic-resistant bacteria. In fact, the American Academy of Pediatrics recommends organic foods to reduce exposure to pesticides, although it is unclear if such a reduction in exposure is clinically relevant (Forman and Silverstein 2012).

A caveat proposed by Reganhold and Wachter (2016), and a debate that will continue, is the nature of nutritionally meaningful differences, and as stated earlier, those that might be deemed negligible in a well-nourished healthy populations. Although the slight differences may not be statistically significant, they could over time have an accumulative affect and advantage in both nutrient content and less exposure to environmental risks and thus a reduction in body burden with an increase in health benefits.

Huber and colleagues (2011) rightly suggested that "the lack of a straightforward relationship between nutritional value and health is another reason why it has been difficult so far to draw conclusions from comparative studies on the health effect of organic food" (p. 7). For example, nutrients and secondary metabolites in plants cannot easily be related to health outcomes. Ingestion, digestion, and absorption are all different factors in understanding the relationship between diet and health. Although still inconclusive, potential health advantages from organic food consumption have been seen through in vitro studies demonstrating higher antioxidative, antimutagenic, and inhibition of cancer cell activity, while animal studies have shown positive effects on fertility indices and the immune system. Epidemiological studies on humans have shown that organic foods may lower risk of allergies, whereas human intervention studies have produced ambiguous results (Huber et al. 2011). Overall, there is no evidence supporting the argument that organic food reduces cancer morbidity and mortality. In fact, the CDC has argued that despite the use of synthetic agrochemicals in conventional farming practices, diets rich in fruits and vegetables have a much lower cancer risk (CDC 1999 in Magkos et al. 2003).

A review by Brantsæter et al. (2017) supports the earlier systematic assessments on the inconclusive positive implications and benefits of organic food on human health (Huber et al.

2011; Forman and Silverstein 2012; Smith-Spangler et al. 2012). One challenge is that few studies are investigating the possible health benefits of organic food consumption in humans. While providing some indications, Brantsæter et al. (2017) suggest that the available evidence is limited and therefore insufficient to conclude whether organic food is healthier than conventional food. Overall, studies report overwhelmingly that there are decreased pesticide levels in organic foods compared to conventional foods.

Consumer Perceptions and Outcomes

Following the Precautionary Principle[6] for environmental health, consuming organic foods is deemed healthy, nutritious, ecological safer, and a socially responsible action. Some environmental health researchers argue that having precautionary actions and policies can help to create opportunities for conducting studies and communicate results to the public in a way that can maintain objectivity while protecting the public health and the environment (Kriebel et al. 2001). Consumer choice of organic foods is driven by factors far beyond nutrients and residues, although this continues to be the focus of research on organic foods. Yet, some limited reviews indicate that consumers beliefs and behaviors are shaped by perceptions of organic food being more environmentally friendly, promoting sustainable farming, protecting small farms, ethically considerate of animal welfare, and healthier and safer for the public health (especially for at-risk populations, including pregnant women, unborn children, the disabled, chronically ill, and elderly) compared to nonorganic conventional foods (Hemmerling et al. 2015; Strassner et al. 2015). However, studies are still lacking a holistic understanding of consumer preferences, behaviors, and attitudes. In fact, some literature suggests that the majority of consumer choices are influenced more by the ethics of safer farming practices than health concerns (Strassner et al. 2015). Interestingly, consumers have an overall higher utility for organic food and therefore have a higher willingness to pay for organics, seen as an investment in the future for their health, community, animals, and environment. In this regard, the literature suggests a need for having a broader "*meta*" view on organics as contributing towards creating a more ethical food system for all—producers, consumers, animals, land, water, air, and environment (Strassner et al. 2015). In turn, it may be these associated factors that in the end support and lead to healthier outcomes.

A report on differing views of the benefits and risk of organic foods in the United States by the Pew Research Center (Funk and Kennedy 2016) shows a link to a growing priority for healthy eating. While a majority of those surveyed saw the health benefits of organic food (although not specified), this opinion was not strongly tied to common divisions such as political lines, education, income, geography or having children, "rather they were tied to individual concerns and philosophies about the relationship between food and well-being" (Funk and Kennedy 2016: 4).

Consumer perceptions' research suggests higher perceived food quality of local organic foods (Canavari and Olson 2007; Carpio and Sengildina-Massa 2009; Hemmerling et al. 2015), increased freshness of products (Roininen et al. 2006), and support of local food production (reducing "food miles"; Denver and Jeansen 2014).

Organic Foods and Healthy Futures

One can take a two-pronged approach for considering the future of organic foods. First, an evidence-based understanding is needed on the comparison between organic and conventional foods. Second, the evidence base, then, will result in not only having a better understanding of the ecology of organic foods but also a better understanding of consumer choices and decision-making surrounding organic foods. We recognize that both must go hand in hand to inform public health policies and interventions.

Regarding evidence-based practices, there are several key considerations for future research on organic food consumption and health outcomes. First, ecohealth perspectives that highlight the cyclical, bidirectional relationship between diet, food systems, environment, and health outcomes, such as those applied globally by the United Nations would be beneficial, indicating that small scale agriculture with less industrial inputs can be more nimble, sustainable, and health promoting. "The organic food system puts the land (agricultural) back into the diet; it is the land from which the diet *in toto* is shaped" (Strassner et al. 2015: 1). To achieve such a goal, we need a restructuring of US agricultural, food, and health policy to directly recognize the link between health and sustainability (Fischer and Garnett 2016).

Second, structural investments in longitudinal, controlled studies to advance the dissemination and implementation of organic food science is necessary in order to understand the complex relationship between health and well-being and our ecology. Specifically, taking a life-course approach is important in an effort to understand exposure, risk, and health outcomes across conception, pregnancy, and from infancy through adulthood. Future comparison between organic and conventional foods must also take into account emerging diet, health, and environment relationships when assessing risk and benefits that are linked to our own intestinal microbiome (Roeselers et al. 2016) and epigenetics outcomes of nutrition and disease (Feinberg and Fallin 2015).

As mentioned previously, to date, there are few controlled trials comparing health outcomes for organic versus conventional foods, thus we rely instead on less accurate cross-sectional epidemiological studies. We are making a call for longitudinal studies that follow cohorts or populations over the long term[7]. This would provide an opportunity to gain a comprehensive understanding of organic food availability, accessibility, affordability, and consumption with health benefits and outcomes. Funding structures should thus focus upon the application of the sustainable diet concept of the FAO (Burlingame and Dernini 2012). Public–private partnerships also need to be nurtured. Just as we have seen with healthier food options for heart diseases and diabetes, this support can be harnessed for understanding organic foods and healthy futures.

Because organic food production is not at the same economy of scale as conventional production, making up less than one percent of total US cropland and less than 5 percent of total US food sales (OTC 2016), there is imperfect knowledge about organic foods. Organic consumers are largely driven by their conscience, considering how their behaviors and patterns of food consumption affect the environment with a deep concern about their own and their families' health and well-being (Funk and Kennedy 2016). It is evident that consumer preferences are informed and shaped by the morality and ethics of farming and perceived benefits linked to animal welfare, environmental consciousness, ecological sustainability, and health benefits. Our discussion on organics is larger than the sum of its parts. It begs us to have a dialogue on the broader political-economic and sociocultural constructions of foodways and their embodied burdens on population well-being. In doing so, we can form an evidence-based recommendations and decisions on organic food and healthy futures.

Notes

1 Economies of scale refers to an increase in production at a scale where production costs decrease, supply increase, and prices decrease, making organic foods more accessible, available, and affordable.

2 Gray literature can be defined as information and scholarship that is not distributed through commercial publishers. This would include, for example, research reports and studies produced, in both print and electronic formats, by government and nongovernmental agencies, academic centers and institutions, and businesses.

3 A *meta-analysis* is a statistical synthesis of the data from separate but similar, comparable studies, leading to a quantitative summary of pooled results. In public health, it is quite standard to "pool" all of the raw data from across the studies that are being analyzed into one "meta-analysis" with the aim to understand the overall data trends using a predetermined criteria of data quality and analysis. Meta-analyses are considered to be highly robust secondary data analysis studies that hold up well to critical appraisal of individual studies by their very nature (Last 2000).

4 Negative externalities is an economic concept that refers to the unintended consequences or negative impacts of agricultural inputs on the environment (Quizhen 2016).

5 Centers for Disease Control and Prevention (CDC) is one of the central operating branches of the Department of Health and Human Services (DHHS) tasked with monitoring America's health status, providing timely and accurate health information, and responding to public health emergencies and natural disasters (CDC 2018).

6 *Precautionary principle* is an environmental health guideline within public health, which promotes the following four central components: taking preventive action in the face of uncertainty; shifting the burden of proof to the proponents of an activity; exploring a wide range of alternatives to possibly harmful actions; and increasing public participation in decision-making (Kriebel et al. 2001).

7 Longitudinal studies collect data on a cohort over a period of time (Last 2000).

References

Annunziata, A., and Vecchio, R. 2016. Organic Farming and Sustainability in Food Choice: An Analysis of Consumer Preference in Southern Italy. *Agriculture and Agricultural Science Procedia* 8: 193–200.

Benbrook, C., Zhao, X., Yáñez, J., Davies, N., and Andrews, P. 2008. *New Evidence Confirms the Nutritional Superiority of Plant-Based Organic Foods*. Washington, DC: The Organic Center. www.organic-center.org/reportfiles/NutrientContentReport.pdf

Benbrook, C., Davis, D. R., and Andrews, P. K. 2009. Methodologic Flaws in Selecting Studies and Comparing Nutrient Concentrations Led Dangour et al. to Miss the Emerging Forest Amid the Trees. *American Journal of Clinical Nutrition* 90: 1700–1701.

Brantsæter, A. L., Ydersbond, T. A., Hoppin, J. A., Haugen, M., and Meltzer, H. M. 2017. Organic Food in the Diet: Exposure and Health Implications. *Annual Review of Public Health* 38: 2.1–2.19.

Burlingame, B., and Dernini, S., eds. 2012. *Sustainable Diets and Biodiversity.* Rome: FAO, Biodiversity International.

Canavari,M., and Olson, K. D., eds. 2007. *Organic Food: Consumers' Choices and Farmers' Opportunities.* New York: Springer.

Carpio, C. E., and Sengildina-Massa, O. 2009. Consumers' Willingness to Pay for Locally Grown Products: The Case of South Carolina. *Agribusiness* 25: 412–26.

Carson, R. 1965. *Sense of Wonder.* New York: Harper and Row.

CDC. 2018. Sustainable Lifestyle. www.cdc.gov/sustainability/lifestyle/index.htm

Dangour, A. D., Allen, E., Lock, K., and Uauy, R. 2009a. Nutritional Quality of Organic Foods: A Systematic Review. *American Journal of Clinical Nutrition* 90: 680–685.

Dangour, A. D., Allen, E., Lock, K., and Uauy, R. 2009b. Reply to DL Gibbon and C Benbrook et al. *American Journal of Clinical Nutrition* 90: 1701.

Delate, K., Cambardella, C., Chase, C., Johanns, A., and Turnbull, R. 2013. The Long-Term Agroecological Research (LTAR) Experiment Supports Organic Yields, Soil Quality, and Economic Performance in Iowa. *Proceedings of the USDA Organic Farming Systems Research Conference.*

Denver, S., and Jeansen, J. D. 2014. Consumer Preferences for Organically and Locally Produced Apples. *Food Quality and Preference* 31: 129–34.

DHHS and USDA. 2015. 2015–2020 Dietary Guidelines for Americans, 8th edn. http://health.gov/dietaryguidelines/2015/guidelines/.

FAO/WHO. 2001. *Codex Alimentarius—Organically Produced Foods.* Rome: Joint FAO/WHO Food Standards Programme, Codex Alimentarius Commission.

Feinberg, A. P., and Fallin, M. D. 2015. Epigenetics at the Crossroads of Genes and the Environment. *Journal of the American Medical Association* 314: 1129–30.

Fischer, C. G., and Garnett, T. 2016. *Plates, Pyramids and Planets: Developments in National Healthy and Sustainable Dietary Guidelines: A State of Play Assessment.* Rome: FAO; Oxford: The Food Climate Research Network at the University of Oxford.

Forman, J., and Silverstein, J. 2012. Organic Foods: Health and Environmental Advantages and Disadvantages. *Pediatrics* 130: e1406–e1415.

Funk, C., and Kennedy, B. 2016. *The New Food Fights: U.S. Public Divides over Food Science. Differing Views on Benefits and Risks of Organic Foods, GMOs as Americans Report Higher Priority for Healthy Eating.* Washington, DC: Pew Research Center.

Gadhoke, P., and Brenton, B. P. 2015. Children in Transition: Visual Methods for Capturing Impressions of Food Landscapes, Family, and Life among Homeless Youth. *Neos: A Publication of the Anthropology of Children and Youth Interest Group* 7: 8–9.

Gadhoke, P., and Brenton, B. P. 2016. Erasure of Indigenous Food Memories and (Re-) imaginations. In *Food Cults: How Fads, Dogma, and Doctrine Influence Diet*, ed. Kima Cargill. Lanham, MD: Rowman and Littlefield.

Gadhoke, P., and Brenton, B. P. 2017. Food Insecurity and Health Disparity Synergisms: Reframing the Praxis of Anthropology and Public Health for Displaced Populations in the United States. In *Responses to Disasters and Climate Change: Understanding Vulnerability and Fostering Resilience*, ed. Michele L. Companion and Miriam S. Chaiken. Raton, FL: CRC Press.

Gibbon, D. L. 2009. Nutrient Content Not a Primary Issue in Choosing to Buy Organic Foods. *American Journal of Clinical Nutrition* 90: 1699–700.

GoGreen. 2015. Guide to Genetically Modified Food. www.gogreen.org/resources/guides/guide-to-genetically-modified-food

Hall, D. C., Baker, B. P., and Jolly, D. A. 1989. Organic Food and Sustainable Agriculture. *Contemporary Policy Issues* 7: 47–72.

Hemmerling, S., Hamm, U., and Spiller, A. 2015. Consumption Behaviour Regarding Organic Food from a Marketing Perspective—A Literature Review. *Organic Agriculture* 5: 277–313.

Huber, M., Rembiałkowska, E., Średnickab, D., Bügel, S., and van de Vijver, L. P. L. 2011. Organic Food and Impact on Human Health: Assessing the Status Quo and Prospects of Research. *Journal of Life Sciences* 58: 103–9.

Kriebel, D., Tickner, J., Epstein, P., Lemons, J., Levins, J., Loechler, E. L., Quinn, M., Rudel, R., Schettler, T., and Stoto, M. 2001. The Precautionary Principle in Environmental Science. *Environmental Health Perspectives* 109: 871–6.

Last, J. M. 2000. *A Dictionary of Epidemiology*, 4th edn. New York: Oxford University Press.

Leifert, C., Rembialkowska, E., Nielson, J. H., Cooper, J. M., and Lueck, L. 2007. Effects of Organic and "Low Input" Production Methods on Food Quality and Safety. *3rd QLIF Congress: Improving Sustainability in Organic and Low Input Food Production Systems.* University of Hohenheim, Germany, 20–23 March.

Magkos, F., Arvaniti, F., and Zampelas, A. 2003. Putting the Safety of Organic Food into Perspective. *Nutrition Research Review* 16: 211–21.

Organic Trade Association. 2016. *U.S. Organic State of the Industry: Market Analysis.* Washington, DC: Organic Trade Association. http://ota.com/sites/default/files/indexed_files/OTA_StateofIndustry_2016.pdf

Reganold, J. P., and Wachter, J. M. 2016. Organic Agriculture in the Twenty-First Century. *Nature Plants* 2: 15221.

Roeselers, G., Bouwman, J., and Levin, E. 2016. The Human Gut Microbiome, Diet, and Health: *Post hoc non ergo propter hoc. Trends in Food Science and Technology* 57: B302–305.

Roininen, K., Arvola, A., and Lahteenmaki, L. 2006. Exploring Consumers' Perceptions of Local Food with Two Different Qualitative Techniques: Laddering and Word Association. *Food Quality and Preference* 17: 20–30.

Smith-Spangler, C., Brandeau, M. L., Hunter, G. E., Bavinger, C., Pearson, M., Eschbach, P. J., Sundaram, V., Liu, H., Schimmer, P., Stave, C., Olkin, I., and Bravata, D. M. 2012. Are Organic Foods Safer or Healthier than Conventional Alternatives? A Systematic Review. *Annals of Internal Medicine* 157: 348–66.

Średnicka-Tober, D., Barański, M., Seal, C. J., Sanderson, R., Benbrook, C., Steinshamn, H., Gromadzka-Ostrowska, J., Rembiałkowska, E., Skwarło-Sońta, K., Eyre, M., Cozzi, G., Larsen, M. K., Jordon, T., Niggli, U., Sakowski, T., Calder, P. C., Burdge, G. C., Sotiraki, S., Stefanakis, A., Stergiadis, S., Yolcu, H., Chatzidimitriou, E., Butler, G., Stewart, G., and Leifert, C. 2016. Higher PUFA and n-3 PUFA, Conjugated Linoleic Acid, α-Tocopherol and Iron, But Lower Iodine and Selenium Concentrations in Organic Milk: A Systematic Literature Review and Meta- and Redundancy Analyses. *British Journal of Nutrition* 115: 1043–60.

Strassner, C., Cavasko, I., Di Cagno, R., Kahl, J., Kesse-Guyot, E., Lairon, D., Lampkin, N., Loes, A.-K., Matt, D., Niggli, U., Paoletti, F., Pehme, S., Rembialkowska, E., Schader, C., and Stolze, M. 2015. How the Organic Food System Supports Sustainable Diets and Translates These into Practices. *Frontiers in Nutrition* 2: 1–6.

Quizhen, C. 2016. *Comparative Study in Agricultural Externalities from Empirical Point of View: Experts' Perspectives, Assessment Levels, and Policy Impacts.* University of Helsinki dissertation.

Worthington, V. 2001. Nutritional Quality of Organic versus Conventional Fruits, Vegetables, and Grains. *Journal of Alternative and Complementary Medcine* 7: 161–73.

Zhao, X., Chambers, E., Matta, Z., Loughin, T. M., and Carey, E. E. 2007. Consumer Sensory Analysis of Organically and Conventionally Grown Vegetables. *Journal of Food Science* 72: S87–91.

11

Profile: Erika Tapp, Concerned Consumer

I can't tell you the number of times I've heard hipsters and community organizers tell me that organic food and urban farming can feed the world. To prove their point, they excitedly reference a pocket park with three raised beds, intermittently tended by *volunteer* high school students from an economically depressed and ethnic part of the city, growing a scraggly assortment of squirrel-ravaged tomatoes and water-deprived greens. "This they say," voices quivering with farm-to-table sanctimosity, "will solve the world food problem, while teaching disadvantaged children the value of eating vegetables!"

Similarly, in Philadelphia I see many urban farms created with the intent of providing food to undernourished communities. Often, they are one-offs by intrepid young hipsters who've taken a weekend course in permaculture and are hell-bent on applying their scanty knowledge in an environment that they perceive as potentially grateful for their middle-class-white intervention. Other times, these mini farms are managed by a combination of paid staff and community volunteers participating in programs funded by not-for-profit donations, but it's rare for these groups to possess sufficient technical expertise to effectively use the growing space.

In my experience in community work, urban farms are excellent tools for community building, neighborhood beautification, giving city kids a chance to turn off screens and play in the dirt, and expand local knowledge of fresh (often organic) produce. I know that there are many excellent schoolyard garden programs that use horticultural practice and vegetable culture to teach academic lessons, but they make no claims to feed the world, only to educate it. The Agatston Urban Nutrition Initiative (in partnership with the Netter Center of the University of Pennsylvania: www.urbannutrition.org/) is a perfect example of the value of using vegetable gardens to teach academic and life lessons, all while making fresh food available to the community. The goal of such programs is often to provide healthy food to what are known as "food deserts," or places with reduced access to whole and fresh foods due to a lack of grocery stores. And while organic is a nice option, and usually promoted by these programs, it's not a requirement for meeting the needs of those who live in food deserts; they simply need better access to a wider range of good quality foods.

The issue with food deserts is food security, of course, and that's what Janet Chrzan's chapter is all about: how organic food can contribute to global food security. Food security isn't about organic vs. conventional, and neither is it guaranteed to provide food for all. Food security depends on availability and accessibility, which isn't technically related to organics; people are food-insecure

FIGURE 11.1 *Bison meat is available at Erika's farmers' market. Photo courtesy of Janet Chrzan.*

FIGURE 11.2 *Abundance in the Global North: The organic market in Paris. Photo courtesy of Janet Chrzan.*

(hungry) because they lack money to buy food. People are hungry because they are poor. Which is a sobering concept since sometimes the privileged assume that local and organic can solve food problems. Local and organic may solve some food problems (and serve to foster a bevy of farm-to-table restaurants) but it's not the solution to feed ten billion in 2050, nor is it the solution to feed the hungry now. In her chapter, Chrzan makes it clear that organics as a solution isn't about choosing organic vs. conventional, but rather fostering agroecological farming practices that can support sustainability through the "Triple Bottom Line": Planet, Profits, People. She convinced me that organics is more than healthy eating and preserving the environment, but also a good way to counter soil degradation and the water and heat stress caused by climate change.

But as she points out with the example of Bhutan, there are values that link food security to organic, by linking soil and agricultural health to population health. From that perspective, making organic food available is the moral and right thing to do; mitigating a food desert involves making sure that healthy food is available, and that might mean organic. From a food justice perspective, it's only right to ensure that the food valued by the privileged (ahem, local, sustainable, organic . . .) is also available to the economically deprived.

There are many organizations that are attempting to do just this, with the goal of making healthy, organic food available in areas not served by reliable markets and stores. Some use a market model to bring food (usually, but not always organic) to areas with a dearth of fresh food. Others are more ambitious, aiming for availability and affordability via community gardens, CSAs, and other similar programs. Here in Philadelphia an excellent example of the former is the Food Trust, which operates farmers' markets (http://thefoodtrust.org/). Since running a market costs money, organizations like the Food Trust use grants and donations to offset the cost of operations. These markets also support use of the SNAP and WIC farmers' market coupons, which allow qualified recipients to receive a monthly allotment of farmers' produce. In addition, many of these farmers' markets (and individual farmers) have Electronic Benefits Transfer (EBT) machines that allow SNAP recipients to buy directly using monthly benefits.

Other programs meld production and access, often using grants and donations to offset costs for those less wealthy. Many CSAs offer sliding rates; one of the better examples is Peacework CSA run by Elizabeth Henderson (www.peaceworkcsa.org). Given that there is often a price premium for organic food, it can be difficult for some to afford the higher prices. Some sort of offset is necessary; a sliding scale is one option, providing that richer CSA members are willing to pay more to support others' access. Alternately, there are CSAs that can accept SNAP payments and vouchers, which is probably one of the more elegant solutions.

Some groups are even more ambitious, blending farming, markets, education, and economic grants to those in need. Mill City Grows (www.millcitygrows.org/), launched by Francey Slater (an alumnus of the Agatston Urban Nutrition Initiative), operates community and schoolyard gardens, an urban farm, markets, and educational programs in Lowell, Massachusetts. Mill City promotes organic gardening and provides produce at half price to qualified recipients. Similarly, Karyn Moskowitz founded New Roots in Louisville, Kentucky (www.newroots.org/) which uses a complicated distribution system to link farmers to markets in underserved areas. Customers reap the benefit of a short distribution chain resulting in lower (and depending on need, often subsidized) food, and farmers benefit from a higher farm gate price than they would receive by selling to distributors. Acting as a link the organization uses grants and donations to offset the costs of bringing farmers and customers together, and explicitly sells in areas without ready access to

FIGURE 11.3 *In the end, it's just about feeding people. Photo courtesy of Leigh Bush.*

farm produce. These organizations take their food justice seriously, but also realize that their model requires economic inputs to survive; they are not sustainable without grants and donations.

And therein lies the problem—producing organic food costs more than farming conventional. That means solving food insecurity with organic food requires economic and political will, and public policies that permit—and make possible—access for all.

Can Organic Farming Feed the World?

Janet Chrzan

Before we go back to organic agriculture in this country, somebody must decide which 50 million Americans we are going to let starve or go hungry. (Earl Butz, then US Secretary of Agriculture, 1971)

Any farmer who shifts to organics only does so because he failed as a conventional, or real, farmer. (Professor of Veterinary Medicine, University of Pennsylvania, 2014)

If everyone shifted to veganism and organics, world hunger would be eliminated. Meat eating and industrial agriculture are the real causes of hunger, not poverty. (Student at the University of Pennsylvania, 2014)

I prefer organic food because it's clean food, it's pure, clean, and won't cause cancer. (Respondent, farmers' market shopper survey, 2016)

Four quotes, all misinformed, but all honestly believed by each person. Their beliefs were channeled and made rational by the information they received during their professional lives, presented as the accepted wisdom of their fields, or, in the case of the latter two, from inaccurate advocacy websites. In all four cases these beliefs were also held by many of their social and professional peers, creating an echo chamber that further valorized the erroneous logic and created a space for absolute certainty.

But each of these quotes is incorrect, to varying degrees, yet something like each can be heard repeated almost daily in the United States, by equally misinformed people, and taken as gospel truth.

According to the UN Food and Agriculture Organization (FAO) and the UN Conference on Trade and Development (UNCTAD) as well as a great many agronomists and agricultural researchers,

Earl Butz's statement is incorrect. According to the very principles and processes of successful organic farming, the professor of veterinary science at the University of Pennsylvania is wrong, because adopting organics is actually harder than most conventional farming practices—it requires more training and education, more planning, and more expertise—organic farming is *knowledge* intensive rather than *input* intensive (Halberg et al. 2006). And unfortunately, changes to farming systems will not solve world hunger; indeed, hunger is caused by poverty in almost all cases. The world currently produces—using mostly conventional farming practices—more than enough food to adequate nourish every single person on earth, 2,870 calories daily, according to the FAO (2015: 24). Hungry people lack *access* to that food; the food is certainly available. And the final quote is doubly problematic, since it belays a fundamental misunderstanding about what organic food is and what it can do—and is informed by fraudulent science about cancer causation.

What might be the connection between hunger and organic food production, if there is one? Would shifting to organics doom a significant number of people to starvation, assuming (perhaps incorrectly) that less food might be available? Or would shifting increase food availability among those who are hungry? Would a shift to organics have different consequences in the Global North than the Global South, and if so, would it change rates of food insecurity in those regions? Can organic food realistically be a part of all diets, not just those of the wealthy? Are there any examples of feeding the hungry using organics that are sustainable, any programs that are working? And finally, can we envision a simultaneously more sustainable, organic future that is also hunger-free? The answer? "It's complicated."

What Is Hunger?

Hunger is an imprecise colloquial term used to denote a variety of situations marked by a lack of food, from an individual tummy growling, to the physical state of being undernourished, to a general lack of food from the household to the nation-state. For this reason, there are several more precise terms designed to identify what kind of hunger a person or groups of persons might be experiencing. "Hunger," on the individual or subjective level, is indeed that uncomfortable feeling that occurs when the stomach is empty. More precisely, according to the World Food Program (WFP) hunger is defined as "not having enough to eat to meet energy requirements" (www.wfp. org/hunger/glossary) and by the USDA as "an individual-level physiological condition that may result from food insecurity" (www.ers.usda.gov/topics/food-nutrition-assistance/food-security-in-the-us/definitions-of-food-security.aspx). Clearly, neither the feeling of hunger nor lack of calories or nutrients has anything to do with organic food directly. Perhaps, then, we should think of "food security," defined as "a situation that exists when all people, at all times, have physical, social and economic access to sufficient, safe and nutritious food that meets their dietary needs and food preferences for an active and healthy life" (FAO World Food Summit 1996). Based on this definition, four food security dimensions can be identified: food availability, economic and physical access to food, food utilization, and stability over time (FAO World Food Summit 1996; FAO 2015: 53). That is a far more complicated concept, and one that can indeed have a connection with the production and intake of organic food.

Food Security is a technical concept that provides variables for measuring access to sufficient food to sustain the biological processes of life. Those variables are complicated but quantifiable, since established metrics can assess the four dimensions on the individual, community, and nation-state levels. For instance, public health monitors and government officials (USDA and ERS officers) in the United States use several measuring protocols to understand food insecurity on different levels. Household food security is queried with eighteen questions designed to establish the frequency of missing meals, among a sampled selection of US households. Households are labeled as having "low food security" if they say "yes" to a certain number of these questions, and will exhibit "very low food security" if they report conditions such as missing meals for a full day because they had no means to access food; these categories also have special considerations for households with children (Coleman-Jensen et al. 2016: 3). In other words, metrics for measuring food insecurity in the United States are designed to approximate the frequency of missing meals, and low food security, or "hunger," indicates that during the year meals are indeed missed.

Food insecurity is measured very differently as the scales enlarge from the domestic household to the nation-state. While there are ways to measure community food security (Cohen 2002; Chaiken et al. 2009), most governments and multilateral organizations use food production estimation tools to predict where food insecurity hotspots may occur. While many of these organization offer online estimations of food security (see the World Food Program's VAM shop; USDA/ERS's "Global Food Security"; FAO's Global Food Security Statistics, IFPRI's Food Security Portal, etc.) almost all use some approximation metric to estimate food needs and availability. For instance, the USDA/ERS uses calories as a basis, and measures potential calorie availability from commodity crops (corn, soybeans, wheat, rice, and cotton) weighting meat and oil production in relation to their caloric equivalents to grains; the calculation of available energy is called a "Food Balance Sheet" (see ERS: www.fao.org/economic/ess/ess-fs/en/). Calculations account for the domestic production of commodities (supply), domestic food utilization (demand), the food supply available for human consumption (minus animal feed), and food exports to arrive at calorie availability per person for that country (FAO 2015; Rosen et al. 2015: 6). From there, using a more complicated calculation that includes income levels and food costs, a distribution gap is calculated to estimate how many people might not receive the benchmark nutritional target of 2,100 calories per day. Obviously, this is a weak estimation, since it fails to consider actual distribution, although there an attempt to approximate income distribution and food access. It also utilizes calories as the target metric, even though complete nutrition requires balanced macronutrients (proteins, carbohydrates, fats, and water) as well as micronutrients (vitamins and minerals). Calories are easier to measure since most agricultural systems aim to produce carbohydrates from commodity crops. Regardless, it is a very broad and imprecise measurement, and is used primarily to predict potential hunger hotspots and to estimate the numbers of people who may, conceptually, be receiving fewer calories than needed. According to the FAO, approximately 11 percent of the global population is undernourished, an estimated 780 million (2015: 8).

Ultimately, the conceptualization and measurement of hunger and food security are in no way related to organics; planning agencies' understanding of hunger focuses solely on production and access to macronutrients. That is not to say that those nutrients are, or are not, produced with organic methods; indeed, there is no reason why they should not be. However, given the widespread perception that organic production is always less per acre than conventional, it is understandable why food security planners generally ignore organics when measuring supply and demand; they see that "X" number of people are hungry, and that agricultural goals must therefore increase production (supply) to meet demand; any use of agricultural processes that might decrease supply must therefore be either ignored or actively discouraged. Mostly, organics are ignored.

The FAO considers food security to be an essential element of the "Right to Food," or one aspect of "Freedom from Want," one of the essential Four Freedoms defined in 1941 by President Franklin Delano Roosevelt, which provide the underlying theoretical goals of the UN Declaration of Human Rights. The Right to Food is achieved "when every man, woman and child, alone or in community with others, has physical and economic access at all times to adequate food or means for its procurement" (Universal Declaration of Human Rights [Article 25], International Covenant on Economic, Social and Cultural Rights [Article 11]). According to the FAO, the Right to Food is a legal concept, because nation-states that have ratified the Article are duty-bearers to fulfill that right. Citizens are rights-holders rather than beneficiaries. Thus, under this article, citizens have the right to adequate food. A related concept is Food Sovereignty, which is defined as "the right of peoples to healthy and culturally appropriate food produced through ecologically sound and sustainable methods, and their right to define their own food and agriculture systems. It puts the aspirations and needs of those who produce, distribute and consume food at the heart of food systems and policies rather than the demands of markets and corporations" (Nyeleni Organization 2007). Food Sovereignty asserts the rights of peoples and nation-states to control the variables that determine their access to food. It is thus a political concept, rather than a legal or technical one. Furthermore, while food sovereignty can serve as a goal and an action plan that could, theoretically, assure food availability, it is not an actionable international or domestic tool, as it has no legal backing or means of enforcement. It does, however, provide a best-case scenario for communities to attain control over their food systems and food economies, and for that reason alone is an important concept.

Yes, but what does this mean in relation to organic food production and intake? Truthfully, not much, which is why it is important to understand these concepts and their limits. How nutritionists, public health planners, and international agencies conceptualize hunger is critical for those who design food policies on any level and with any goal, including those designed to increase organic production. For instance, if nation-state planners desire to increase food availability (one of the four pillars of food security) and they genuinely believe, as did Mr Butz, that organic production results in less food than conventional, they may intentionally or unintentionally prejudice food policy away from organics and towards conventional. Similarly, if community planners are convinced that organic food is safe, pure and clean (as did the farmers' market shopper) and thus a better choice for public health, they may support agricultural policies that privilege organic production. And while the Right to Food does not specify how food is farmed (only that it is adequate for nutritional and biological needs), if the right to food is articulated with a strong Food Sovereignty focus, the community may decide that organic production is preferred to provide "healthy and culturally appropriate food produced through ecologically sound and sustainable methods." And that is precisely the choice made by Bhutan, which has pledged to go 100 percent organic by 2020 in order to increase farmer productivity, boost incomes and preserve the environment (Neuhoff et al. 2014).

Clearly, within these various statutes, policies, and declarations there are items which could be construed to favor adoption of organics as a core food policy, and, as Bhutan acknowledges, may indeed be preferred if they are designed to increase farmer income and protect the environment. And therein lies the deeper and far more complex analysis of how organics may relate to world hunger. First, return to the statistic that the world currently makes roughly 2,870 kilocalories (Kcals, also simply called "calories") per person daily (FAO 2015: 24). The first requirement of food security is met, therefore—that of availability or production. The problem is that economic and physical access to food, the second element of food security, is curtailed. Obviously, availability may differ by region for a number of reasons, but in the most basic sense, and for those regions not experiencing active warfare or famine, there is food enough for all. Access is almost always determined by economics; this was illustrated by Mark Bittman:

> "Put yourself in the poorest place you can think of … Now, are you hungry? Are you going to go hungry? Are you going to have a problem finding food?" The answer, obviously, is "no." Because almost all of you … would be standing in that country with some $20 bills and a wallet filled with credit cards. And you would go buy yourself something to eat. The difference between you and the hungry is not production levels; it's money. There are no hungry people with money; there isn't a shortage of food, nor is there a distribution problem. There is an I-don't-have-the-land-and-resources-to-produce-my-own-food, nor-can-I-afford-to-buy-food problem. (Bittman 2014)

The solution is obvious: raise the incomes of the poor to a level sufficient to buy enough food. It's that simple. If there is a market, the food will flow towards it. Unfortunately, as we all know, it's not that simple in the absence of a universal income provision, in part because of the last sentence in Bittman's quote—lack of access to land and resources to grow food or work sufficient hours to buy food. Compounding the difficulty is that the vast majority of the hungry poor are also farmers or farm laborers of some sort, especially in the Global South. In low-income countries, 48 percent of the population is in poverty (living on less than $1.25 per day) and 28 percent is undernourished as of 2010. Furthermore, 78 percent of the poor were rural (900 million people), and 750 million of them were working in agriculture (63 percent of the global poor; Townsend 2015). So the very people who grow food are often the ones who lack it (two excellent books on why this occurs are Thurow and Kilman 2009 and Thurow 2013). The solution, then, is to raise the incomes of the poor, and especially of the rural, agricultural poor, while ensuring adequate harvests to feed the world population.

Can Organic Farming Increase Farmers' Incomes?

Raising farm incomes has been one focus of development planning for a very long time; the US land grant universities were started, in part, to assure farmer incomes, production and sustainability. Post–World War II development efforts included income-increasing programs (such as import substitution) as well as commodity crop production enhancement (such as the Green Revolution technologies). Too often development efforts focused on encouraging cash cropping and commodity production with green revolution technology without focusing on the need to explicitly increase the availability of food crops (Goldman 1998; Serageldin 1998). To quote Goldman, "Food poverty alleviation must be understood as a process of income growth, but income growth in many countries must be understood as a process of agricultural, and most directly, food production growth" (1998: 299). This is in contrast to development schemes designed to uphold economic goals for nation-state comparative advantage within a global food system and which encourage the production of non–food commodity crops, such as hemp or sugar cane. Accordingly, the economic anthropologist Keith Hart has argued that organic agriculture has the potential to raise rural farm incomes because the price premium received allows farmers to realize a profit and thus create the positive feedback loop that fuels a symbiotic and economically equal relationship between urban industries and rural food production (Hart 2004). Using James Steuart's (1767) thesis that creating a mutually advantageous rural/urban exchange of goods, services, and foods ensures wealth for both areas, Hart argues that the organic price premium primes this development pump, raising food prices enough that farmers can afford urban goods and increasing the market—and thus urban employment and factory profits—sufficiently to support economic mutuality. But the key to this exchange is higher food prices and farm profitability, not a cheap food policy.

And yes indeed, organic products command a higher farm gate price in both the developed and developing world. Certified as well as non-certified organic crops achieve a price premium when sold via wholesale, short-chain, and direct-to-consumer (see the USDA ERS price comparison tables: www.ers.usda.gov/data-products/organic-prices.aspx as well as the USDA analysis of prices: www.ers.usda.gov/topics/natural-resources-environment/organic-agriculture/organic-market-overview.aspx). A recent meta-analysis demonstrated that global prices for organic crops were 29–32 percent higher than conventional and, even with higher labor and distribution costs, tended to provide a 22–35 percent greater profit margin (Crowder and Reganold 2014). Similar findings for profitability were also noted in a 2009 FAO study that analyzed different crops and farming variables in over fifty studies, demonstrating that organic farms were more frequently profitable than conventional and far more profitable during stressful growing conditions, such as drought (Nemes 2009; see also Pretty 2002; Pretty et al. 2002; Delate et al. 2003; Duram 2005: 42–50; Pimental et al. 2005; Halberg et al. 2006; Rodale Institute 2011; Willer and Lermoud 2016). Research also indicates that organic farm systems can provide a boost in economic development and prosperity for municipalities and regions (UNEP-UNCTAD 2008; Jaenicke 2016; Reganold and Wachter 2016), which supports Hart's thesis and could increase incomes outside of the farm sector as well.

However, readers may wonder how much of this profitability requires the current price premium, and what would happen to the farmgate price if supply of organic food increased sharply? Presumably, once supply and demand equalized, the prices could, theoretically, drop down to a point where reduction in farm output would lead to a loss of profitability. Crowder and Reganold (2015) and Reganold and Wachter (2016) provide a meta-analysis that demonstrates 5–7 percent greater profitability even when organic price premiums are removed. The Rodale 30-Year Organic Trial (Pimental et al. 2005; Rodale 2011) also demonstrated farm profitability regardless of premiums. Part of the reason for increased profits—among those farmers who have mastered the intricacies of successful organic growing and who have survived the transition period—is a reduced need for external inputs that decreases farmer debt levels and financial risk. In other words, production costs are lower, while production levels may be the same or even higher, in some cases. Pretty et al. (2002) linked lower costs to improved water efficiency, increased soil health leading to greater production fertility, and decreased pesticide costs. The Rodale Trial also measured decreases in oil and gas costs and savings with a shift to no-till and other "efficient" systems, and others have documented decreased costs for fertilizers and other inputs. Nemes's (2009: 10–13) meta-analysis found that variable costs were often lower (fertilizers, pesticides, plowing, tillage, etc.) but fixed costs (administrative, land rent/purchase, etc.) similar to conventional systems; the increased labor costs and needs were significant, although the combination of higher yields, higher prices, and decreased input costs allowed for profitability even with increased labor. While organic farming requires 35 percent more labor (Pimental et al. 2005: 14), actual labor costs can be spread out over the growing season rather than requiring hired, and often expensive, additional labor at planting and harvesting, resulting in an average increase in labor costs of only 15 percent. In addition, the timing of tasks required for organic farming are more amenable to the family labor systems common in developing regions. Reasons for profitability, especially in the developing world context, are higher yields, few financial outlays, market access and premium farmgate prices, fewer price fluctuations, and the potential for increased farm prices overall as sufficient organic saturation can lead to higher farm prices in general (Parrott et al. 2006: 167–8).

The higher labor costs are most likely to affect farmers in developed nations, where labor costs more and where farmers may have to rely on hired (rather than family) labor. In developing world systems and among peasant farmers, the use of family members can offset increased labor costs. The reduction of inputs necessary for production in an established organic farm that uses crop rotation, agroecological, and other on-farm methods for fertility enhancement is also ideal for developing world farming systems in which farmers have little ready cash for input purchase. And finally, positive externalities such as environmental/ecosystem services can boost production levels (such as the creation of environments supportive of beneficial animals and insects) and further increase yields, although measuring external inputs is difficult (Reganold and Wachter 2016). The outcomes can increase farm profit, increase farm family income, and thus increase food security.

Can Organic Farming Increase Farmers' Yields?

But can organic farming really produce equal or greater yields than conventional farming systems? And here again, the answer seems to be "it's complicated" and it depends on where the farms are located, which processes have been used for growing, and what crops are grown. In the developed world, where conventional farmers utilize a full array of fertilizers, pesticides, machinery and technologies to boost crop yields, some crops farmed organically demonstrate a decrease in yield initially. On the other hand, in the developing world where minimal inputs may be available, shifting to an organic system can boost yields dramatically. Critical questions remain, however: can first-world organic farming systems increase yields over time to rival those of industrial agriculture? And can third-world systems sustain increased outputs given the vagaries of weather, availability of inputs, and needs of the population? Unfortunately, there has not been as much research conducted on organic systems as on conventional, so while there are indications of yield trends, there are too few studies that provide long-term data on changes in yield over time.

Some recent meta-analyses of organic vs. conventional yields have found that organic systems result in decreased yields. De Ponti et al. (2012) compared a number of published studies to find that organics produces roughly 80 percent of conventional yields, although with much variation between crops, context and scale. Seufort et al. (2012) also found decreased yields, with a range from 5 to 34 percent, with the latter occurring when systems were most similar and optimal, such as corn grown in Kansas. Their study examined context, however, and concluded that many crops, grown organically with good management practices, could almost approach the yields of conventional agriculture and that when crops are ecologically stressed, the yield disparity drops; they call for additional research to better understand the contextualizing factors that determine outputs. Similarly, Badgely et al. (2007), also using a meta-analysis, found that yields averaged 8 percent lower, with organic yields less than conventional in the developed world but, on average, at parity or greater in the developing world. Other studies have found disparities from 20 to 6 percent, depending on crop and context (Halweil 2006); one long-term comparative study conducted in Nebraska found corn, soy, and sorghum yields to be reduced in organic cropping in relation to conventional by 6–33 percent, but that wheat yields were greater by 10 percent (Wortman et al. 2012). John Reganold (2017) summarizes these studies to point out that the National Research Council (2010) found that "that sufficient productivity is only one of four main goals that must be met for agriculture to be sustainable. The other three are enhancing the natural resource base and environment, making farming financially viable, and contributing to the well-being of farmers and their communities."

Other studies have found that organics produces a greater yield than conventional. Christos Vasilikiotis (2005), Pretty et al. (2002), Halberg et al. (2006), and others have noted either near parity or increased yields, depending on crop rotations, environmental conditions, and time since conversion; Duram provides an extensive analysis of studies of comparison yields (2005: 42–50).

Perhaps the most persuasive trial in the United States is that of the Rodale Institute, which compared five-year crop rotations of corn, soybeans and legumes using conventional and organic farming systems. Their twenty-year report (Pimental et al. 2005: 8–10) found that after the first five-year transition period corn yields are at parity, soybeans almost the same, but legumes less for organic systems. However, corn yields during drought years were between 28 and 34 percent greater for the organic fields, and soybean yields dramatically higher (100 percent) in extreme drought years. The summation thirty-year report (Rodale 2011) found that over the course of the full trial corn, soybean, and wheat yields were equivalent but that during drought years, organic corn yields were 31 percent higher than conventional, and that both corn and soybeans tolerated higher weed field infiltration but produced similar yields, indicating that far less weed suppression chemicals could be applied if farmers switched to organic field techniques. Multiyear analyses suggest that with proper legume rotation and other advanced organic farming systems, yields can approach conventional without increasing the agricultural land base (Delate and Cambardella 2004; Badgeley et al. 2007).

Results are far more dramatic in the developing world. The UNEP-UNCTAD report *Organic Agriculture and Food Security in Africa* (2008) compared cropping systems to find that eleven out of thirteen examples demonstrated greater yields with organic processes, with all food crop trials showing increased yields. Organic and near-organic (sustainable) practices in 114 trials in twenty-four countries resulted in average increased yields of 100 percent with the additional benefit of better soil quality, water retention, and greater drought resistance. Similar findings were reported by Pretty et al. (2006) when comparing 286 sustainable agriculture projects in fifty-seven countries; the average crop yield increase was 79 percent with improved water efficiency, and crops showed water use efficiency gains, decreased pesticide use and increased carbon sequestration. And a comparative study of forty projects commissioned by the UK's Governmental

Office for Science (Foresight 2011) reached similar conclusions. However, it must be pointed out that studies of organic systems in many developing nations are not the same as those that study organic farm yields of IFOAM-regulated organic standards as found in the developed world; these studies often examine agroecological farming using sustainable principles rather than farms that adhere to IFOAM organic principles. In effect, they are comparing farming systems that embrace agroecological and sustainable practices rather than those which adhere strictly to a certification system; organics in these cases might be thought of as a big-tent policy rather than a regulated practice. Freyer et al. (2015) provide a superb analysis of the use of organic IFOAM principles, contrasting high-input systems (conventional, often Global North) with what they term low-input (organic, agroecological, often in the developing world) to increase food security; they maintain that it is the values embedded within the principles that can best guide how these systems are evaluated, and how adoption of organic standards can serve to increase food security.

To examine yield data with a food security (rather than an agronomy) focus, it seems that yield disparities are greater between conventional and organic systems that employ high-input processes, especially those for commodity crops within the developed Global North. For farming systems in the developing world in which minimal inputs may have been the norm due to economic or resource constraints, use of organic and/or agroecological methods can dramatically boost outputs, especially for food crops. Furthermore, among high-input commodity crops, conversion to organic creates a decrease in yield, which is to be expected given that such crops are currently gown under ideal conditions, provided with maximum inputs, and bred for the express purpose, usually, of maximum yield. Farming the modern corn/soy rotation is a highly optimized, technological process; deviation from that process is bound to create deficits in output. However, as the USDA crop data demonstrates, adoption of organic techniques for other food crops (such as vegetables, legumes, etc.) can increase yields once conversion is complete, albeit at increased labor cost (for a full examination of these trade-offs, see Halberg et al. 2006), and the Rodale Trial suggests that once organic systems are in stable production, yields increase to levels similar to conventional. In other words, while yield of commodity crops in the Global North (or developed world) may decrease with initial conversion, in the long term yield parity will potentially and probably occur. Furthermore, yields of grains, legumes, and staple and commodity crops can potentially increase in the developing world. In effect, the cautious answer might be that while some crops may decrease, others might be able to fill the gap, and if targeted research increases knowledge of agroecological systems, these deficits might be overcome entirely without increasing the use of arable land.

Can Organic Farming Feed the Future?

The world is facing profound food security challenges. Population is growing rapidly, and expected to crest at approximately nine to ten billion in 2050. Climate change is decreasing yields in many parts of the world due to drought, extreme weather, and shifting crop production patterns. Increasing desire for meat in developing and middle-income countries is creating widespread demand for grains to feed animals, and continued utilization of biofuels in the Global North is diverting grain production from food use to the internal combustion engine. In the face of these problems, several solutions have been suggested. Frances Moore Lappe's seminal book *Diet for a Small Planet* (1971) proposed that vegetarian diets could feed more people because the conversion ratio from grain to meat makes meat ecologically and economically expensive. While this is accurate and a nice idea, unfortunately as nations (and the people within them) become more wealthy, they tend to prefer to eat more meat, not less. Thus the demographics of global development are in direct opposition to the economics of ecology and food security. Which does not mean, of course, that shifting to less meat is not a very good idea (for both personal health and global food security) but it probably is not going to solve the world food problem. Biofuels may or may not be here to stay; certainly in the face of a food crisis affecting the Global North there would be a shift away from using grain for gasoline. But we cannot ignore population growth and climate change, and their effects on demand. And it is the predictive models of how climate change will affect food security that have suggested to agronomists that a shift to organic and sustainable farming practices could increase food access and yields long term, especially for smallholder farmers of the developing world.

Recent meta-analyses, especially those provided by the FAO and other agencies of the UN, have strongly suggested that a shift to organic/sustainable farming is not only preferable but necessary for future food security, particularly in the Global South. It is the externalized costs of conventional, combined with the externalized benefits of organic farming that have suggested this solution. The FAO observes that 90 percent of increased production will need to come from intensification of farming, rather than increased land use, particularly in developing countries (FAO 2009), but this intensification must also preserve natural resources and soil quality; simply dumping more fertilizer on crops will not provide for sustainable gains in production, since the growth rates of cereal yields have been slowing over the last few decades (FAO 2009: 20; Ray et al. 2013). The natural resource base must be preserved in order to ensure environmental services necessary to protect future harvests. The FAO recommends that technological options must both increase productivity and sustainably manage natural resources: "conservation agriculture, avoiding deforestation, forest conservation and management, agroforestry for food or energy, land restoration, recovery of biogas and waste and, in general, a wide set of strategies that promote the conservation of soil and water resources by improving their quality, availability and efficiency of use" (FAO 2009: 30). In 2016 the FAO defined these goals forcefully by stating that increasing cereal yields was possible only by conservation agriculture, particularly for smallholder farmers in the Global South:

FAO's approach to sustainable crop production intensification is the "Save and Grow" model. Save and Grow promotes a productive agriculture that conserves and enhances natural resources. It uses an ecosystem approach that draws on nature's contribution to crop growth, such as soil organic matter, water flow regulation, pollination and natural predation of pests. It applies appropriate external inputs at the right time and in the right amount to improved crop varieties that are resilient to climate change and use nutrients, water and external inputs more efficiently. Increasing resource use efficiency, cutting the use of fossil fuels and reducing direct environmental degradation are key components of the approach, saving money for farmers and preventing the negative effects of overusing particular inputs. (FAO 2016: 50)

The FAO calls for intensified, agroecological farming: "There is no single blueprint for Save and Grow and its ecosystem-based approach to crop production intensification. No magic seeds or technologies exist that will improve the social, economic and environmental performance of cereal production across all landscapes, in all regions. Save and Grow represents a major shift *from a homogenous model of crop production to knowledge-intensive, often location-specific, farming systems"* (Reeves et al. 2016: 98). These shifts in practice are also supported by Worldwatch and the Natural Research Council of the United States, which recommend that farming practices including building soil fertility, agroforestry, urban farming, green manure, improving water conservation and preserving biodiversity, could mitigate climate change; these are methods often used in organic farming (National Research Council 2010; Reynolds and Nierenberg 2012).

In particular, there is the need to support agricultural systems that both mitigate climate change and increase production, and sustainable, agroecological farming can contribute to those goals better than currant conventional farming systems. Decreasing greenhouse gas (GHG) emissions is a critical step in mitigation, and shifting to organics can sequester carbon as well:

> A minimum scenario of conversion to organic farming would mitigate no less than 40 per cent of the world's agricultural GHG emissions. When combining organic farming with reduced tillage techniques under the optimum scenario, the sequestration rates on arable land could easily be increased to 500 kg of carbon/ha per year. This optimum organic scenario would mitigate 4 Gt CO_2-eq per year or 65 per cent of agricultural GHGs. Another approximately 20 per cent of agricultural GHGs could be reduced by abandoning the use of industrially produced nitrogen fertilizers, as is practiced by organic farms. As a result, organic agriculture could become almost climate neutral. (Hoffman, 2010: 19)

The FAO has labeled this shift the "Climate Smart Approach," with three goals: "sustainably increasing agricultural productivity to support equitable increases in incomes, food security and development; increasing adaptive capacity and resilience to shocks at multiple levels, from farm to national; and reducing greenhouse gas emissions and increasing carbon sequestration where possible" (FAO 2016: 14). Shifting to organic farming in Africa furthers these goals, particularly if coupled with support for more efficient markets for crops and value-added farm products (see Auerbach et al. 2013 and Reeves et al. 2016 for case studies of how organic farming can fulfill these goals in developing nations). These practices are easily adopted by smallholder farmers, and are also in their economic and cultural interests (Holt-Gimenez 2017). Above all, this approach provides a Food Security focus, since it actively supports income growth to build farmer resilience, increased yields to provide food for both farm families and for sale, and mitigation services to ensure that agricultural practices can withstand the potential problems of climate change. Acknowledging that current practices in conventional agriculture contribute to greenhouse gas emissions, waste water, deplete soil fertility and fail to guarantee a sustainable livelihood for farmers worldwide, multilateral agencies across the globe are calling for a shift to sustainable and organic farming practices to ensure sufficient food for the future.

Conclusion

Ultimately, ensuring food security for the global population is technically simple, since all that is needed is sufficient cash and/or capacities to produce food and support for the research necessary to improve farming processes. But it is also terribly difficult, because those solutions require political will to ensure effective and fair markets, sufficient payment for crops produced, and financial support for agroecological research. People are hungry because they are poor, even though we currently produce food enough for all. Paradoxically, often the poorest and most hungry are those who produce food. In the future our growing population will require increased food production, which must occur even though climate change, soil erosion, water loss and diminishing growth rates of conventional cereal yields all point to the potential for deep deficits in supply in the face of sharp increases in demand. Facing such trials, food production must not only meet needs but mitigate climate and ecological problems; farmers must grow enough food to provide for themselves, their communities, and the world in a manner that is sustainable, equitable, and protective of the economy, health and the environment. In business theory, this is thought of as the "Triple Bottom Line," sometimes paraphrased as "People, Planet, Profit" to indicate that the system must nourish all three in order to remain functional, sustainable, and ultimately, reliably profitable. Farmers must be able to make enough money to support their families and contribute to their communities; they must grow enough food to feed themselves and the world, and they must do so in a manner which does not diminish the capacity for production of future generations. This chapter hopefully makes clear that organic farming, sustainable farming and/or agroecology is a potential solution to these needs, a solution supported by a wide variety of global agronomists and multilateral agencies. Rather than diminishing food production, organic and sustainable farming can potentially increase outputs, especially under the hotter and drier conditions expected due to climate change. Central to this potential is a shift from an industrial view of agriculture to what is termed the agrarian philosophy of agriculture—one that views agriculture as having an important social function beyond its ability to produce food and other products (Thompson 2010). The goals for farm and food sustainability are precisely those of an agrarian philosophy, since they support the productivity, quality, affordability and accessibility of farming systems, improve soil, air, and water quality and biodiversity, potentially increase the economic viability of farms and farmers and enhance the quality of life within farming communities (National Research Council 2010: 26). These goals could ensure Food Security, might be able to support Food Sovereignty, and can contribute to the fulfillment of the Right to Food, the three concepts that best define adequate food for all. Above all, organic and agro-ecological farming systems can increase farm and community resilience, which will be critical to ensure food security for a population of ten billion in 2050.

References

Auerbach, R., Rundgren, G., and Scialabba, N. E. 2013. *Organic Agriculture: African Experiences in Resilience and Sustainability*. Rome: FAO.

Badgley, C., Moghtader, J., Quintero, E., Zakem, E., Chappell, J., Aviles-Vazquez, K. K., Samulon, A., and Ivette Perfecto. 2007. Organic Agriculture and the Global Food Supply. *Renewable Agriculture and Food Systems* 22: 86–108.

Bittman, M. 2014. Don't Ask How to Feed the 9 Billion. *New York Times*. November 11, sec. Opinion. http://nyti.ms/1Er9MfX.

Chaiken, M., Dixon, J. R., Powers, C., and Wetzler, E. 2009. Asking the Right Questions: Community-Based Strategies to Combat Hunger. *NAPA Bulletin* 32: 42–54.

Cohen, B. 2002. *Community Food Security Assessment Toolkit*. Washington, DC: US Department of Agriculture, Economic Research Service.

Coleman-Jensen, A., Rabbitt, M., Gregory, C. A., and Singh, A. 2016. Household Food Security in the United States in 2015. Washington, DC: US Department of Agriculture, Economic Research Service. www.ers.usda.gov/webdocs/publications/err215/err-215.pdf.

Crowder, D., and Reganold, J. 2015. Financial Competitiveness of Organic Agriculture on a Global Scale. *Proceedings of the National Academy of Sciences* 112: 7611–16.

Delate, K., and Cambardella, C. A. 2004. Agroecosystem Performance during Transition to Certified Organic Grain Production. *Agronomy Journal* 96: 1288–98.

Delate, K., Duffy, M., Chase, C., Holste, A., Friedrich, H., and Wantate, N. 2003. An Economic Comparison of Organic and Conventional Grain Crops in a Long-Term Agroecological Research (LTAR) Site in Lowa. *American Journal of Alternative Agriculture* 18: 59–69.

Duram, L. 2005. *Good Growing: Why Organic Farming Works. Vol. 17. Our Sustainable Future.* Lincoln: University of Nebraska Press.

FAO. 1996. Declaration on World Food Security and World Food Summit Plan of Action. World Food Summit, November 13–17, 1996.

FAO. 2009. *How to Feed the World in 2050. High Level Expert Forum.* Rome: FAO. www.fao.org/fileadmin/templates/wsfs/docs/expert_paper/How_to_Feed_the_World_in_2050.pdf.

FAO. 2015. *FAO Statistical Pocketbook World Food and Agriculture 2015.* Rome: FAO.

FAO. 2016. *The State of Food Insecurity in the World.* Rome: FAO.

Foresight. 2011. *The Future of Food and Farming.* London: The Government Office for Science. https://www.gov.uk/government/uploads/system/uploads/attachment_data/file/288088/11-547-future-of-food-and-farming-summary.pdf.

Freyer, B., Bingen, J., Klimek, M. and Paxton, R. 2015. Feeding the World—The Contributions of IFOAM Principles. In *Rethinking Organic Food and Farming in a Changing World*, 81–106. New York: Springer.

Goldman, R. 1998. Food and Food Poverty: Perspectives on Distribution. *Social Research* 66: 283–304.

Halberg, N., Sulser, T., Hogh-Jensen, H., Rosegrant, M., and Knudsen, M. T. 2006. The Impact of Organic Farming on Food Security in a Regional and Global Persective. In *Global Development of Organic Agriculture: Challenges and Prospects*, 277–322. Oxfordshire, UK: CABI Publishing.

Halweil, B. 2006. Can Organic Farming Feed Us All? *World Watch Magazine* 19: 18–24.

Hart, K. 2004. The Political Economy of Food in an Unequal World. In *Politics of Food*, 199–220. Oxford: Berg.

Hoffmann, U. 2010. *Assuring Food Security in Developing Countries under the Challenges of Climate Change: Key Trade and Development Issues of a Fundamental Transformation of Agriculture.* UNCTAD Discussion Paper. Geneva: United Nations.

Holt-Gimenez, E. 2017. Peasants, Science, and Climate Change. *Huffington Post.* 4–11. www.huffingtonpost.com/entry/peasants-science-and-climate-change_us_58ed7a79e4b081da6ad008f6.

Jaenicke, E. 2016. *U.S. Organic Hotspots and Their Benefit to Local Economies.* Washington, DC: Organic Trade Association.

Lappe, F. M. 1971. *Diet for a Small Planet.* New York: Ballentine Books.

National Research Council, Committee on Twenty-First Century Systems Agriculture. 2010. *Toward Sustainable Agricultural Systems in the 21st Century.* Washington, DC: National Academies Press.

Nemes, N. 2009. *Comparative Analysis of Organic and Non-Organic Farming Systems: A Critical Assessment of Farm Profitability.* Rome: FAO.

Neuhoff, D., Tashi, S., Rahmann, G., and Denich, M. 2014. Organic Agriculture in Bhutan: Potential and Challenges. *Organic Agriculture* 4: 209–21.

Nyeleni Organization. 2007. Declaration of Nyeleni. https://nyeleni.org/spip.php?page=NWarticle.en&id_article=375.

Parrott, N., Olesen, J. E., and Hogh-Jensen, H. 2006. Certified and Non-Certified Organic Farming in the Developing World. In *Global Development of Organic Agriculture: Challenges and Prospects*, 153–79. Wallingford: CABI Publishing.

Pimental, D., Hepperly, P., Hanson, J., Douds, D., and Seidel, R. 2005. Environmental, Energetic, and Economic Comparisons of Organic and Conventional Farming Systems. *Bioscience* 55: 573–82.

Pimental, D., Hepperly, P., Hanson, J., Seidel, R., and Douds, D. 2005. *Organic and Conventional Farming Systems: Environmental and Economic Issues*. Kutztown, PA: The Rodale Institute.

Ponti, T., Rijk, B., and van Ittersum, M. K. 2012. The Crop Yield Gap between Organic and Conventional Agriculture. *Agricultural Systems* 108: 1–9.

Pretty, J. N. 2002. Lessons from Certified and Non Certified Organic Projects in Developing Countries. In *Organic Agriculture, Environment and Food Security*, ed. N. Scialabba and C. Hattam. Rome: FAO.

Pretty, J. N., Morison, J. I. L., and Hine, R. E. 2002. Reducing Food Poverty by Increasing Agricultural Sustainability in Developing Countries. *Agriculture, Ecosystems and Environment* 95: 217–34.

Pretty, J. N., Noble, A. D., Bossio, D., Dixon, J., Hine, R. E., Penning de Vries, F. W. T., and Morison, J. I. L. 2006. Resource-Conserving Agriculture Increases Yields in Developing Countries. *Environmental Science and Technology* 40: 1114–19.

Ray, D. K., Mueller, N. D., West, P. C. and Foley, J. A. 2013. Yield Trends Are Insufficient to Double Global Crop Production by 2050. *PLOS One* 8: 1–8.

Reeves, T., Thomas, G., and Ransay, G. 2016. *Maize, Rice Wheat: A Guide to Sustainable Cereal Production*. Rome: FAO.

Reganold, J. P. 2017. Comparing Apples with Oranges. *Nature* 485: 176.

Reganold, J. P., and Wachter, J. P. 2016. Organic Agriculture in the Twenty-First Century. *Nature Plants* 2: 1–8.

Reynolds, L., and Danielle, N. 2012. *Innovations in Sustainable Agriculture: Supporting Climate-Friendly Food Production*. Washington, DC: Worldwatch Institute.

Rodale Institute. 2011. *The Farming Systems Trial: Celebrating 30 Years*. Kutztown, PA: Rodale Institute.

Rosen, S., Meade, B., and Murray, A. 2015. *International Food Security Assessment, 2015–2025*. Washington, DC: US Department of Agriculture, Economic Research Service. www.ers.usda.gov/webdocs/publications/gfa26/53198_gfa26.pdf?v=42184.

Serageldin, I. 1998. Sustainable Agriculture for a Food Secure World. *Social Research* 66: 105–16.

Seufert, V., Ramankutty, N., and Foley, J. 2012. Comparing the Yields of Organic and Conventional Agriculture. *Nature* 485: 229–32.

Steuart, J. 1767. *Principles of Political Oeconomy (Two Volumes)*. London: Miller and Caddell.

Thompson, P. B. n.d. *The Agrarian Vision: Sustainability and Environmental Ethics*. Lexington, KY: University of Kentucky Press.

Thurow, R. 2013. *The Last Hunger Season: A Year in an African Farm Community on the Brink of Change*. New York: PublicAffairs.

Thurow, R., and Kilman, S. 2009. *Enough: Why the World's Poorest Starve in an Age of Plenty*. New York: PublicAffairs.

Townsend, R. 2015. *Ending Poverty and Hunger by 2030: An Agenda for the Global Food System*. Washington, DC: The World Bank. http://documents.worldbank.org/curated/en/700061468334490682/Ending-poverty-and-hunger-by-2030-an-agenda-for-the-global-food-system.

UN International Covenant on Economic, Social and Cultural Rights, Article 11 § (1966). www.ohchr.org/EN/ProfessionalInterest/Pages/CESCR.aspx.

UNEP-UNCTAD Capacity Building Task Force on Trade, Environment and Development. 2008. *Organic Agriculture and Food Security in Africa*. Rome: United Nations.

Vasilikiotis, C. 2005. Can Organic Farming "Feed the World"? http://agroeco.org/doc/organic_feed_world.pdf.

Willer, H., and Lermoud, J. 2016. *The World of Organic Agriculture: Statistics and Emerging Trends 2016*. Bonn: IFOAM–Organics International.

Wortman, S., Galusha, T., Mason, S., and Francis, C. 2012. Soil Fertility and Crop Yields in Long-Term Organic and Conventional Cropping Systems in Eastern Nebraska. *Renewable Agriculture and Food Systems* 27: 200–216.

12

Profile: Leigh Bush, Food Anthropologist

I study food; technically I'm getting a Ph.D. in studying food. But I'd have to admit that it wasn't until my second year in graduate school that I actually bothered to grow my own food. I was ashamed of my ignorance of the life cycle; so, after a heart-wrenching breakup, when I needed some healthy distraction, I decided to take advantage of the large backyard I had at my disposal and plant a garden. Tilling up sod, turning over rich black soil—that was the easy part—but now I was just one sad girl with one large dark patch of ground and no clue what to do next.

I headed to Bloomington's magnificent farmers' market to try to buy a clue. I had been to the market dozens of times in the past—buying produce as it came in throughout the seasons. But that day was the first time I had thought about backtracking my seasonal knowledge from eating, to the life cycles of preparing, planting, growing, and harvesting. When I found the seed stand I was mesmerized by the varieties I had to choose from: lettuces, tomatoes, beans, potatoes, the list went on. The farmer not only knew who had collected each seed and from what farm but was also able to tell me which tasted the best, which were easy to grow, which produced in bounty but bolted as soon as the weather got hot, etc. She told me how to amend my soil, how to plant each variety of seed, and when and how to water so I didn't shock my young seedlings. I couldn't believe I had passed this stand at the market so many times without realizing the wealth of knowledge residing there.

I took my seeds home and started planting, and watering, and waiting, and watching. It felt like a miracle when the plants began to sprout from the ground. I was proud, but it didn't take long for the weeds and pests to discover my bounty too. Not to mention, in my heartbroken ambition, I had planted a garden twice as big as I needed. Luckily, I was able to call upon my friends for a garden day, which we spent under the sun, sweaty, and with dirt under our fingernails and caked into our pores as we dug up dandelions and dusted off aphids. My friends told me about partner plants, natural pest deterrents, and why ladybugs are as helpful as they are adorable. Sharing our knowledge about the living earth that was literally in my backyard made me feel empowered not to conquer, but to collaborate with the land and all it had to offer. Moreover, for the first time since I had arrived to school, I felt a family gathering around me, like I was making a home.

As the month passed and my sprouts sped toward the sun, I decided to have a celebration honoring all the help I had received along the way. I would make a feast, including a huge salad featuring all the greens from my backyard, which I would cut fresh for the occasion. The evening

FIGURE 12.1 *Leigh picking berries. Photo courtesy of Leigh Bush.*

FIGURE 12.2 *Leigh with a slug. Photo courtesy of Leigh Bush.*

FIGURE 12.3 *Leigh and her friends in the garden. Photo courtesy of Leigh Bush.*

came and, as usual, I was running late. With the guests already arriving, I hurriedly ventured out to the backyard, scissors in hand, to harvest my bounty of greens. Only when I bent over the first bunch, something was wrong. The leaves were covered in slugs dining on my delicious produce! Frantically I cut leaf after leaf only to turn up more slugs. I was devastated that my leaves were not only half-eaten, but I didn't have enough time or the equipment to rinse off all the slug goo. I was in tears for my failure—so much work and help but I had nothing to offer in return. Again, my community came to my rescue. Within twenty minutes the grapevine was activated and a few people arrived with their own local lettuce harvests. Another couple arrived with a sack of eggshells from their backyard chickens, showing me how crushed shells around the base of a plant could keep the slugs from climbing up the stalks. Together we tossed together the multi-sourced salad to match my rabbit stew and roasted broccoli. With sad-turned-happy tears still fresh in my eyes, I toasted my thoughtful, caring, and collaborative community that had come together to make this meal happen.

This chapter raises important questions about what we eat, why we eat it, how it's grown, and what it all means. It makes clear that interconnectedness has determined human history—regardless of spiritual belief systems—and that losing our awareness of interconnection has had

FIGURE 12.4 *Garden fresh food! Photo courtesy of Leigh Bush.*

tangible negative repercussions on our environment and our ability to feed ourselves. How was it that we somehow lost our sense of this connection? Was it the "the chemicalization of food," as Dr. Robbins asserts? And how did we develop the parallel association that both soil and dirt are "bad"? Was it because so few of us farm anymore? Did the lack of connection to this rich and nourishing substrate, soil, also lead us to desacralize our food?

And how has a resistance to the loss of food's sacredness been embedded in the organics movement? The early "organics" mission integrated spiritual, physical, ecological and psychological health, stressing the connectedness of people, the soil, their land and their food. This concept, known as anthroposophy, is a wholistic/holistic/integrative approach to food, farming, and human relations. Unfortunately, modern organic food has fused with corporate capitalism, and this may have resulted in a belief that eating organic is more about individual identity and "health, freshness, and taste," rather than an interconnected food system. This has also led to identifying organic food as what it is NOT rather than what it is; it is not "supposed" to be full of chemicals; but is that the sum total of what organic can be?

As you read this chapter you might think, like I did, that humus is a funny word, but that it's also undeniably connected to humans and humanity. My experience growing food put that understanding back into perspective for me. What started out as a singular and lonely endeavor to create life from scratch and on my own turned into an experience that integrated me with my local community and brought me closer to my neighbors and friends. In creating humus I found not just dirt but also the foundations of humanity, which are in interconnection, patience, knowledge sharing, and compassion. Organics "transcendental" value isn't just about mystifying dirt, but rather, organics is about how the physical act of growing and eating healthy food functions to establish similarly healthy relationships with our environments and communities. Local agriculture, with its focus on these relationships and communities, may be one way to reestablish the original "transcendental" approach to organics as a social movement; as people build relationships with farmers, their land and their food they can better understand the process of making, repairing, and maintain humus, our living human/humanity life force.

The Transcendental Meanings
of Organic Food

Richard H. Robbins

For it is as a living thing, not as dead medium, that the soil is most important to us.
—LORD NORTHBOURNE

Dirt, Food, and Meaning

To begin to understand the meaning of organic foods it is best to begin with dirt. For much of recorded history human beings viewed dirt (or soil) with reverence. In Gen. 3.19 we are told that

> By the sweat of your brow you will eat your food until you return to the ground, since from it you were taken; for dust you are and to dust you will return.[1]

The human creation story continues in the same vein: "Adam" in Hebrew means "soil," and "Eve," who emerges from Adam means "life." And in Latin "soil" (humus) and "human" (homo) come from the same Indo-European root meaning "earth" (Descola 2010: xv).

In Mt. 13.2–9, the Parable of the Sower expressed soil's generative power:

> A farmer went out to sow his seed. As he was scattering the seed, some fell along the path, and the birds came and ate it up. Some fell on rocky places, where it did not have much soil … Still other seed fell on good soil, where it produced a crop—a hundred, sixty or thirty times what was sown.

Fertility rituals throughout history and in societies all over the world, testify to the special relationship between humans and soil. In the middle ages, farmers evoked God as the creator and keeper of the soil: they sprinkled it with holy water prior to planting, or erected or buried a crucifix in the soil, or planted the first seeds in the shape of a cross. People worshipped Mary as a symbol of the living soil out of which Christ emerged (Patzel 2010: 261ff). Beginning in the early nineteenth century, thousands of pilgrims came to the Catholic shrine of El Santuario de Chimayó in New Mexico to collect soil that, when mixed with water, is said to have healing powers when eaten or rubbed on the skin (Rossbacher and Rhodes 2007). At modern Catholic churches in central Africa, farmers bring soil to be blessed by the priests to ensure fertility (Aguilar 2002). Anthropologist Carol Delaney (1991) describes how in traditional Turkish villages producing children is analogous to the planting and growing of crops; the man provides the "seed" with his semen, and the woman serves as the "soil" in which the seed germinates and grows (see also Anderson 2013: 189–90).

Somewhere along the line, the special meanings of soil began to change. It may have been Justus von Liebig's reduction of necessary plant nutrients to nitrogen, phosphorus, and potassium, and his critique of humus-based theories of plant growth. It may have been the emergence of new technologies, the mechanization of agriculture, or the mass commodification of food first in England in the eighteenth century or in the United States in the nineteenth. Regardless, whatever special relationship believed to exist between the farmer and the soil underwent a radical transformation. John Steinbeck in *Grapes of Wrath* (1939: 47–8) captured the change well. Sitting in his tractor, he wrote, the modern farmer "could not see the land as it was, he could not smell the land as it smelled; his feet did not stamp the clods or feel the warmth and power of the earth."

> He sat in an iron seat and stepped on iron pedals ... He did not know or own or trust or beseech the land. If a seed dropped did not germinate, it was no skin off his ass. If the young thrusting plant withered in drought or drowned in a flood of rain, it was no more to the driver than to the tractor.
>
> He loved the land no more than the bank loved the land ...
>
> And when that crop grew, and was harvested, no man had crumbled a hot clod in his fingers and let the earth sift past his fingertips. No man had touched the seed, or lusted for the growth. Men ate what they had not raised, had no connection with the bread. The land bore under iron, and under iron gradually died; for it was not loved or hated, it had no prayers or curses.

Our language usage reflects the change in how we view the soil; synonyms for "dirt" include "crud," "grime," "gunk," "muck," while antonyms include "clean" and "cleanliness." We speak of things being "soiled," getting a "dirty deal," being "dirt poor," or someone as a "dirtbag," and so on. Sacred soil has been transformed into profane dirt.

The desacralization of soil parallels the historical desacralization of food sources, in general. Corn, heavily ritualized in indigenous cultures of South and North America (see, e.g., Vogt 1993) is now largely food for livestock, an industrial byproduct, fuel, or a food additive. Animals, such as cows, chickens and pigs, once enshrined with sacred meanings as bulls, cocks, and boars (see Ohnuki-Tierney 1990; Pollan 2006; Lawler 2016), like "dirt," are now used as negative, often sexist or homophobic, epithets.

Professional food theorists, says Mary Douglas (1984: 5), recognize this change:

> In the sense that the whole of society has been secularized by the withdrawal of specialized activities from a religious framework, so food has been taken out of any common metaphysical scheme. Moral and social symbols seem to have been drained from its use.

However, she says, it is a mistake to overemphasize the exclusive concern with food as physical nourishment and ignore what she calls "the ordinary consuming public in modern industrial society," who, as she says, "works hard to invest its food with moral, social, and aesthetic meanings" (Douglas 1984: 5; see also Douglas 1972).

In fact, resistance to the desacralization of food and agriculture associated with the industrialization of the food supply and the chemicalization of agriculture extends well back into the nineteenth century, beginning with Sylvester Graham's (1794–1851) promotion of vegetarianism, temperance, and whole grains. The effort to instill social meanings into our food continues today with various food movements including vegetarianism, Fair Trade, direct farmer-to-consumer marketing, the Slow Food movement and especially, the organic food movement (see, e.g., Guthman 2003, 2004; Belasco 2007; Fromartz 2007).

Most of these movements embody what we might characterize as *transcendental* meanings that imbue food and/or the agricultural process and world in general with spirituality or some variant of secular or *naturalized* spirituality (Solomon 2003: 25) that contrasts generally with the *mechanical* world view that underlies the modern scientific understanding of natural processes. The organic food movement represents one example of this protest.

Social movements, however, change. They shift scope, membership, and even goals, but generally some core values remain. *We want to examine the extent to which the organic agriculture and food movement of today retains the original meanings, goals and values of those who founded the movement.* We could do this by examining the research on the relative benefits of organic relative to conventional food and agriculture and why and who chooses organic. Most of this research has asked whether one type of food is more nutritious than the other, whether it produces greater yields, is healthier, more environmentally sustainable, friendlier to local farmers or lower in cost. But these studies produce inconclusive results. Most studies on yield find organically grown food somewhat lower than conventional (see Badgley and Perfecto 2007; Seufert et al. 2012). Yet these studies ignore the fact that we have chosen to commercially produce, as Anderson (2013: 185–6) notes, only a relatively small number of available foods— wheat, rice, maize, potatoes, beans, cattle, sheep, goats, pigs, and chickens, sugar, coffee, tea, and chocolate—all relatively easy to grow, process, and store and that fit in well with an industrial farming model. Nutritionally, organically produced food is at least as nutritious, if not more so, than conventional food (see, e.g., Worthington 2001; Barański et al. 2014; Średnicka-Tober et al. 2016). And as the scale of organic production increases, and as the negative externalities of industrial agriculture are included in their costs, the price of organic begins to approach that of industrial food.

To a large extent, arguments over yields, costs, and nutrition of industrial and organic foods respectively miss the point. Food has meaning; what, where, and how a person eats represents a statement about their view of the world. That does not mean the differences are not important; effects on the land, on the body, and on the environment are real. But eating, from an anthropological perspective, is a grand ritual; it is performative, the eater or the food producer acting out a view of the world, and, in so doing, making that view all the more real. It begins with food choice, but extends to where it is bought, how it is cooked, with whom it is eaten, and even how it is chewed (see Bourdeau 1984: 190–91).

As we mentioned above, the organic movement was all about dirt, or more specifically, "humus"—the organic component of soil created by the decomposition of leaves and other plant material by soil microorganisms. Founders of the movement such as Rudolf Steiner, Sir Albert Howard, Lord Northbourne, and Lady Eve Balfour, whose ideas we will examine below, were responding to the chemicalization of agriculture promoted by the discoveries of Justus von Liebig (1803–1873). Liebig, generally considered the founder of modern chemistry, argued that all that was needed for crop and plant growth were the proper amounts of nitrogen, phosphorous, and potassium, specifically dismissing humus as a necessary ingredient in agriculture. When Fritz Haber (1868–1934) invented a process to transform nitrogen gas into ammonia in 1909, thus inventing fertilizer (along with the gases that killed and disabled tens of thousands in World War I), he, along with Liebig, became the founders of modern industrial or chemical agriculture.

Von Liebig and other Enlightenment thinkers construed the universe as a machine, and the scientist (a term which first appeared in 1834) sought laws regarding the relationships between things in order to manipulate those relationships. The proper study of nature involves constructing a correct model from experience and making it a neutral object of study (Taylor 1989). Within this framework "nature," the essential object of study has no meaning beyond its function and utility, with a value dependent only on how it can be controlled and manipulated (Hanna Schösler et al. 2013: 443).

The massive shift to industrial farming and the world view on which it was based, to which the founders of organic methods responded, did not occur overnight and did not really overwhelm traditional farming in acreage until after World War II. Its characteristics—the use of synthetic chemical fertilizers, pesticides, the use of genetically modified organisms, concentrated animal feeding operations, heavy irrigation, intensive tillage and concentrated monoculture production—were advertised to increase yields and profit, although yield increase did not generally occur until after the so-called Green Revolution with its heavy investment of water and energy resources. Yet as the shift occurred one must assume that there were other factors involved in the choice of industrial farming and that we need to understand that choice as well as the one to farm or eat organically. Certainly corporate profit, reduction of labor input, the absence of accountability for negative externalities (e.g., pollution, water extraction, etc.), and government incentives played major roles in the transformation. But, we would argue, it had also to do with philosophical outlooks, how one viewed nature. Understanding why farmers choose to grow their crops and raise their livestock organically, and why consumers choose to buy organic products requires an understanding of the meanings embedded in both a mechanical and transcendental view of nature.

In the remainder of this chapter we will examine the ideas of the founders of the humus or organic food movement and then examine the extent to which their ideas inform the choices of both farmers and consumers today and what this means for the future of organics.

The Founders: Steiner, Howard, Northbourne, and Balfour

While many people had a role in the development and expansion of organic farming, Rudolf Steiner (1861–1925), Sir Albert Howard (1873–1947), Lord Northbourne (1896–1982), and Lady Eve Balfour

(1898–1990) were among the most prominent in defining its philosophy, purpose, and meaning. They shared at least five characteristics:

- An emphasis on the soil, its health and maintenance.

- A spiritual, sometimes religious and transcendental orientation to nature.

- An appreciation of indigenous and "peasant" agricultural practices, particularly those of India and China.

- A strong motivation for social reform and resistance to technology and the growing industrialization of food production and society in general.

- A belief that farming methods reflected the state of society and the physical and psychological health of its members.

While their approach differed from that of industrial farming advocates, they nevertheless claimed that their ideas were based on sound scientific principles.

Rudolf Steiner's agricultural philosophy or "bio-dynamic farming," emerged from his mystical intuitions; the farmer, he said, must recapture the "astral-ethereal forces" on his land (Peters 1979: 20–21; Conford 2001). His Christian mysticism that he called "Anthroposophy" constituted an attempt to unite the material and spiritual understandings of human beings. Science, he said, separated agriculture from deeper essences, and failed to understand the farm as a "living organism" (Steiner 1924; Peters 1979: 27).

Steiner's agricultural philosophy contained magical elements. He viewed the soil as alive, not only with living organisms, but as a force. For example, he advised farmers to fill a cow horn with manure and bury it over the winter in order to capture the "astral-ethereal forces" on his land (Peters 1979: 29), a practice reminiscent of garden magic all over the world (see, e.g., Malinowski 1935). Biodynamic farmers avoided chemicals, and tried "to improve the health and vitality of the soil" by working with "the health-bearing forces of nature." In healthy soil "seeds will bring forth plants which are true to their own unique nature and have more life-giving vitality to offer animals and humans" (see Fromartz 2007: 11).

Steiner, like the advocates of organic farming who followed him, appreciated indigenous agricultural practitioners: "I have always considered what the peasants and farmers thought about things," Steiner (1924a) said, "far wiser than what the scientists were thinking."

Sir Albert Howard also emphasized the value of studying indigenous methods. The scientist studying agriculture, said Howard, must become a "brother cultivator" with the peasant; only then would the "new investigator" succeed where, in Howard's eyes, past researchers had failed (Peters 1979: 44). Howard developed much of his organic philosophy with his work in India. He served as England's agricultural advisor to the Indian government and Director of the Institute of Plant Industry in Indore, where, along with his wife, Gabrielle Louise Caroline Matthaei, and her sister Louise (whom he married after Gabrielle's death in 1930) they developed the "Indore Method" of composting, still central to organic farming and home gardening.

Howard, like Steiner, saw the soil in general, and humus specifically, as the focus of his work. Liebig, who claimed that humus was unimportant in plant growth, was wrong, said Howard.

The benefits of humus might be indirect, but they could improve yields and health for animals and crops.

In his writings and practices, Howard mixed scientific and religious metaphors. In his major works, *An Agricultural Testament* (1940) and *The Soil and Health* (1947), he spoke of a pantheistic respect for "Mother Earth," for the "Wheel of Life," for "God's Green Carpet," and for "Nature's Law of Return" (Peters 1979).

While Howard did not share the magical elements in Steiner's philosophy, he did share Steiner's view that the meaning of humus farming was not only about farming methods, but about the state of nature and the state of society itself. Liebig's method of chemical fertilizers, Howard said, "is based on a complete misconception of plant nutrition. It is superficial and fundamentally unsound ... Artificial manures lead inevitably to *artificial nutrition, artificial food, artificial animals, and finally to artificial men and women*" (Howard 1943: 174, 1947: 228, italics added; see also Fromartz 2007: 9).

Like Howard, Steiner advocated social reform, seeing his agricultural work as just one part of the societal overhaul outlined in his book, *Toward Social Renewal: Rethinking the Basis of Society* (1919).

Lord Northbourne's (Walter James) book, *Look to the Land* (2003), is another work central to defining the meaning of organic agriculture. Northbourne was the first to use the term "organic farming," contrasting it to what he called "chemical farming." He argued that the "chemicalization" of the food chain is "far more deleterious than has yet become clear." Challenging the "artificial manure industry" would be difficult, Northbourne said.

> But we may have to relearn how to treat the land before we can manage entirely without them, or without poisonous sprays ... imported chemicals can by no means make up for a loss of biological self-sufficiency. (2003: 103)

The goal of conquering nature he said, makes no sense; "The idea of conquering nature is as sensible as if a man should try to cut off his own head so as to isolate his superior faculties" (Northbourne 2003: 191).

Northbourne (2003: 1) shared with the other founders of humus or organic farming a deeply spiritual nature and the idea that our approach to the soil defines who we are:

> There is a very real economic and biological linkage, comprehensive and of infinite complexity, between all living creatures ... That which links together all the phases and processes of farming, and which is therefore its foundation, is the soil in all its infinite variety. If the soil is the foundation of farming, the soil is also the foundation of the physical life of man. It is the background of his life. (p. 3)

Northbourne also wrote about the ill-effects of eating what he called "fearsome, tainted, bleached, washed out, and long-dead material" that passed for conventional food and resulted in relatively "poor people" eating too much. Foreseeing the emergence of the local food movement, he said:

> It is ludicrous to cart stuff about all over the world, so someone can make a "profit" out of doing so, when that stuff could much better be produced where it is wanted. (Northbourne 2003: 104)

Northbourne was well ahead of his time in one other respect—the influence of economics, finance, and especially debt on the growth of industrial farming. This debt, a feature of a "peculiar economic system," he said, is unpayable, the burden of interest alone being repressive. The only way to produce the interest is to sell more and more, so, as he put it, "purely financial considerations have everywhere acquired dominance over all others" (Northbourne 2003: 14; Paull 2014; see also DiMuzio and Robbins 2016).

While the debts can never be repaid, he says, they require new technologies and the rationalizing of production as the necessary solution to repayment of debt, rationalizing implying "merely an increase of speed," so that "the same number of men should produce more in a given time" (Northbourne 2003: 16; see also Robbins 2005).

Lady Eve Balfour quoted extensively from Northbourne in her book, *The Living Soil* (1943). And like Steiner, Howard and Northbourne, claimed that organic farming had implications that went well beyond food and farming: "healthy soil meant healthy plants that meant healthy people." At the age of twenty-one, Balfour used her inheritance to purchase a large farm in Suffolk and launched what became known as the Haughley Experiment, the first long-term, side-by-side comparison of organic and chemical-based farming (Balfour 1976).

In sum, we see that the interests of the founders involved more than asking about comparative yields, nutrition, local economies, and so forth; they were concerned largely with the relationship between food and farming and society in general. They espoused a transcendental vision the core of which consisted of how food and farming defined our place in nature and the universe. The question we need to address next is the extent to which the visions of the founders of humus or organic agriculture inform the modern organic agriculture and food movement and what this can tell us about the future of organic food.

Modern Organic Food

The founders embraced an agriculture and food policy that explicitly rejected the technology, politics, and, in the case of Northbourne (and probably Steiner), the economic system of which it was a part. The emergent dominant ideology against which they railed was bad for nature, it was bad for people, and it was bad for society.

Humus-based farming and its ideological foundation may have remained a footnote in food history had it not been for three developments. First, in 1942 J. I. Rodale began popularizing humus-based or organic gardening and agriculture first in his publication *Organic Farming*, then *Organic Gardening and Farming*, today called *Organic Gardening*. Rodale also warned readers against DDT, but the message did not resonate with voters until 1962, when Rachel Carson published *Silent Spring*. Carson's indictment of the chemical industry created popular support for a national Environmental Protection Agency and a ban against using DDT on crops. Finally, the counterculture movements of the 1960s and 1970s adopted many of the ideas of the founders, creating, as Warren Belasco put it (1997: 193), a "countercuisine" movement: "Like earlier dietary missionaries," he said, protestors "believed that the battle for the stomach would determine the fate of the world."

With these developments, beginning in the 1980s, interest in organics soared. The movement became so successful that from their beginnings in the early 1980s a small, organic food store,

Safer Way Natural Foods in Austin, Texas, evolved into the food giant Whole Foods, and a 2.5-acre raspberry farm became the corporate giant Earthbound Farms. Organics hit the big time, consumed, at least sometimes, by 80 percent of the population with 80 percent produced by giant, corporate producers (see Guthman 2003, 2004; Shapin 2006; Fromertz 2007).

The question is what happened to the meaning of organic that had emerged from the concern for soil? Has the organic movement retained that transcendental frame? Part of the answer lies in the demographics of organic food consumers.

Most studies trying to identify the organic food consumer use standard demographic categories (gender, income, age, ethnicity, location, etc.) and come to markedly different conclusions. Some identify high users among older, religious males and females (Bellows et al. 2010); others find increased likelihood of purchasing organic vegetables as levels of income and education increase (Dettman 2008), while another concludes:

> The profile of an organic user is most likely to consist of an Hispanic household residing in the Western United States with children under 6 years old and a household head older than 54 with at least a college degree. (Smith et al. 2009: 731; see also McKenzie et al. 2013)

A Gallop Poll in 2013 concluded that a little less than half of Americans, 45 percent, actively try to include organic foods in their diets, while 15 percent actively avoid them. More than a third, 38 percent, say they "don't think either way" about organic foods. Those who actively try to include organic foods in their diets were likely to be younger (18–29), politically identify as a Democrat, or have an annual household income of $75,000 or more (Riffkin 2014). The US Organic Trade Association concluded in 2014 that the demographic profile (i.e., age, gender, income, education, ethnicity, etc.) of US organic consumers matched the demographic of the country as a whole; in other words, using standard demographic characteristics, there is little to distinguish the organic consumer from the nonorganic consumer, the same finding made by the Hartman Group as early as 1999 (Fromertz 2007: 246).

There is some evidence, however, that the transcendental worldview persists in at least a small proportion of organic shoppers (see, e.g., Chrzan 2016).

One study (Hamilton et al. 1995: 497) randomly interviewed 400 visitors to the annual Mind, Body and Spirit Festival held annually in London to determine the extent to which following an alternative diet expressed a desire for a "spiritual dimension in daily existence," and whether "this type of dietary practice might be seen as a form of implicit religion or at least implicitly to express a religious orientation."

They found that 74 percent were following unorthodox diets, 60 percent were vegetarians, and 45 percent were following organic diets. There was a strong correlation also with the involvement in alternative therapies (e.g., aromatherapy, acupuncture, herbalism, Shiatsu, etc.), involvement with human potential and related groups (e.g., encounter groups, transcendental meditation, etc.), and support for green and environmental groups. They conclude that:

> alternative dietary practices are to some extent associated with a world view of which ecological concerns, holistic orientations and perhaps seekership are a prominent part. (Hamilton et al. 1995: 508)

Other research (Schösler et al. 2013) on Dutch citizens also found that the historical values of the founders characterize those of current organic consumers, and that these values are also shared by a larger part of Dutch society not consuming organic foods.

Emphasizing that we need to understand the cultural context of the choice to buy organic, the study sought to understand the "food philosophy" of individuals, that is the "cluster of practices, values and beliefs" that are "shared on a collective level" (Schösler et al. 2013: 440–41). Their findings were based on in-depth interviews with thirteen people contacted in organic food shops or who mentioned that they regularly used organic foods. The participants described being connected to nature, being part of it, which created feelings of care and responsibility for animals and concerns for one's physical and mental health. They associate food with a sense of well-being and happiness. They also expressed a concern for purity in their food because people "don't know what [they] are eating because food producers mix substances together and thereby obscure people's choices" (p. 450). These attitudes, they say, are characteristic of a "romantic" worldview (or transcendental, as we have called it) that contrasts with a more enlightenment or mechanistic world view. Participants also emphasized the desire to live a moral life. They conclude that, if one examines the central elements of organic food philosophy—the value of connectedness to nature, the relationship between awareness and wellness, the transparency of moral aspects of food choices, and supporting norms that value temperance—are values supported by Dutch people in general.

Yet, in spite of the research on British and Dutch citizens discussed above, it is hard to square the view that our food philosophy is aligning with that of the organic founders when, as one consumer advocate told Michael Pollan, "Organic is becoming what we hoped it would be an alternative to." As Steve Shapin (2006) notes:

> For many who participated in the early phase of organic farming, its subsequent history is a story of paradise lost—or, worse, sold—in which cherished ideals have simply become part of the sales pitch.

Today consumers as well as organic certifiers, at least in the United States, generally identify organic food by what it is not, rather than by what it is; no pesticides, no synthetic fertilizers, no sewage sludge, no genetically modified organisms, and no ionizing radiation. There is nothing in certification guidelines about how organic is more sustainable, or uses less water, energy or other resources, all central goals of the movement. Furthermore, Steiner, Howard, Northbourne, and Balfour emphasized community building, as did those who promoted a countercuisine in the 1960s and 1970s; they emphasized the relationships of people to each other, to nature, and to the universe in general. While these views remain among some segments of our society, they are not likely to characterize the vast majority of those who choose organic food. Individual health, freshness and taste, the most common reasons people give for choosing organic, evidence an individualistic or even ego-centric posture to food (see, e.g., Magnusson et al. 2003; Shephard et al. 2005; Kareklas et al. 2014). This, then, raises the question of whether the organic food movement, as it was originally envisioned, has ceased to exist, coopted instead by commercial interests to maintain or increase profitability? Or has the movement shifted its focus to more local concerns, a shift evidenced perhaps by the dramatic growth in the United States of farmers' markets, community-supported agriculture, and farm to table distributional schemes?

The Local Food Alternative

The organic food and agriculture movement began as a reaction to the increasing dominance of chemical agriculture and the rise of corporate farming. Its goal of preserving the soil, which came to symbolize for the founders the state of the environment and society itself, became secondary for consumers relative to personal health, fresh food, and cost. Given the increasing urbanization of the population, it should not be surprising that the soil lost its salience as a potent symbol. This is not to say that it is forgotten (see, e.g., Logan 2007 [1995]; Shiva 2008; Ohlson 2014), but humus is rarely mentioned in common discourse about food.

As a practical matter, however, humus is critical for plant growth, including the grasses that feed livestock. And where concern and knowledge remains high is among farmers.

As Chrzan (2010: 84) points out, for farmers "organic is a process." And whether a farm is certified organic or not, it is difficult to convey to consumers the farming practice.

The question then is: how can its original meaning of "organic" be retained when the food industry, consumers, government, and nongovernmental agencies no longer understand it as a process, but rather see food as a "thing" (see Chrzan 2010)?

The answer may be the promotion of local food. Local agriculture was never far from the center of concern for Steiner, Howard, Northbourne and Balfour. They often framed their concern about local agriculture around issues of food security.

Amory Starr (2010) argues that local food fits the criteria of a social movement that represents a new emphasis on social relations of trust into the food system; as Chrzan (2010) emphasizes, visiting the source of one's food, exposes consumers to farming stories in which food becomes part of a process as opposed to a thing. Furthermore, and perhaps most importantly, it restores the importance of farmers themselves, transforming them into "active teachers, discussing agronomy, varieties, and cooking" (Starr 2010: 485; see also Berry 1977).

Starr cites the growth in the number of farms reported in the 2007 US Census of agriculture, especially small farms, as possible evidence of renewed interest in farming, especially among the young. And while this growth contracted in the 2012 Census, that may be attributable to the global economic contraction of 2008 and a decrease in so-called *hobby farms*, created largely for tax purposes.

Additional evidence of the interest in local agriculture is the global growth of urban faming and gardens, particularly in low-income communities. Estimates of the percentage global food production in urban areas range from 15 to 20 percent. The more that people become involved with the growing and raising of food, the closer they get to the dirt. But, in addition, such cooperative endeavors also accomplish one of the missions of the founders; that is building community through food. As Starr (2010: 484) puts it:

> Food is transformed from a commodity to a pleasure made possible by human relationships, the limitations/specificities of an ecology, attentive husbandry of biodiversity, and responsible global citizenship.

Conclusion

The organic or humus movement began as a reaction to chemical or industrial agriculture and the substitution of nitrogen, phosphorus and potassium for humus as the medium for plant growth.

The founders of the movement—Rudolf Steiner, Sir Albert Howard, Lord Northbourne, and Lady Eve Balfour—promoted organic farming and food as a solution to the ills of growing industrialization symbolized, for them, by the chemicalization of agriculture. Their movement gained impetus in the 1960s prompted by fears of environmental devastation and the growth of the 1960s counterculture that adopted organic food and culture as one of its symbols of change. The increase in availability of organic food and its cooptation by large food corporations in the 1980s and 1990s, however, has been accompanied by the decrease in whatever transcendental meaning organic agriculture and food held for the founders. The question is, has organic food become only a marketing label to allow corporate agriculture to increase the profitability of its products? If so, to the extent that what we eat influences what we are, food loses the transcendental qualities that have marked it as a significant symbol throughout human history, and leaves it to marketers to impose meanings that serve only to sell more of it. In that case, one solution may be the promotion of local food systems, either rural or urban, that bring people back to where our food is produced and into contact with those who produce it and who appreciate the role and importance of soil in our lives.

Note

1 Although the Hebrew word *afar* is translated as "dust," dirt is probably a more accurate translation.

References

Aguilar, M. 2002. Postcolonial African Theology in Kabasele Lumbala. *Theological Studies* 63: 302–23.

Anderson, E. N. 2013. Food Cultures: Linking People to Landscapes. In *Nature and Culture: Rebuilding Lost Connections*, ed. Sarah Pilgrim and Jules N. Pretty. London: Earthscan.

Badgley, C., and Perfecto, I. 2007. Can Organic Agriculture Feed the World? *Renewable Agriculture and Food Systems* 22: 80–85.

Balfour, E. 1976. *The Living Soil and the Haughley Experiment*. Basingstoke: Palgrave Macmillan.

Balfour, E. 2007 (1943). *The Living Soil*. London: Soil Association.

Barański, M., Średnicka-Tober, D., Volakakis, N., Seal, C., Sanderson, R., Stewart, G. B., Benbrook, C., Biavati, B., Markellou, E., Giotis, C., Gromadzka-Ostrowska, J., Rembiałkowska, E., Skwarło-Sońta, K., Tahvonen, R., Janovska, D., Niggli, U., Nicot, P., and Leifert, C. 2014. Higher Antioxidant and Lower Cadmium Concentrations and Lower Incidence of Pesticide Residues in Organically Grown Crops: A Systematic Literature Review and Meta-analyses. *British Journal of Nutrition* 112: 794–811.

Barry, W. 1977. *The Unsettling of America: Culture and Agriculture*. San Francisco: Sierra Club Books.

Belasco, W. 1997. Food, Morality and Social Reform. In *Morality and Health*, ed. Allan M. Brandt and Paul Rozin. New York: Routledge.

Belasco, W. 2006 (1989). *Appetite for Change: How the Counterculture Took on the Food Industry*, 2nd edn. Ithaca, NY: Cornell University Press.

Bellows, A. C., Alcaraz, G., and Hallman, W. K. 2010. Gender and Food, A Study of Attitudes in the USA towards Organic, Local, U.S. Grown, and GM-Free Foods. *Appetite* 55: 540–50.

Bourdieu, P. 1984. *Distinction: A Social Critique of the Judgment of Taste*. Translated by Richard Nice. Cambridge, MA: Harvard University Press.

Chrzan, J. 2010. The American Omnivore's Dilemma: Who Constructs Organic Food? *Food and Foodways* 18: 81–95.

Chrzan, J. 2016. Organics: Food, Fantasy or Fetish? Paper presented at the annual meeting of the Association for the Study of Food and Society, Toronto, June 24.

Conford, P. 2001. *The Origins of the Organic Movement*. Edinburgh: Floris Books.

Delaney, C. 1991. *The Seed and the Soil: Gender and Cosmology in Turkish Village Society*. Berkeley: University of California Press.

Dettmann, R. L. 2008. *Organic Produce: Who's Eating It? A Demographic Profile of Organic Produce Consumers*. Selected paper prepared for presentation at the American Agricultural Economics Association annual meeting, Orlando, FL, July 27–29, 2008. https://ageconsearch.umn.edu/bitstream/6446/2/467595.pdf

DiMuzio, T., and Robbins, R. H. 2016. *Debt as Power*. Manchester: Manchester University Press.

Douglas, M. 1972. Deciphering a Meal. *Daedalus* 101: 61–81.

Douglas, M. 1984. Standard Social Uses of Food: Introduction. In *Food in the Social Order: Studies of Food and Festivities in Three American Communities*, ed. Mary Douglas. London: Routledge.

Fromartz, S. 2007. *Organic, Inc.: Natural Foods and How They Grew*. New York: Harvest Books.

Guthmann, J. 2003. Fast Food/Organic Food: Reflexive Tastes and the Making of "Yuppie Chow." Social and Cultural Geography 4: 45–58.

Guthmann, J. 2004. *Agrarian Dreams: The Paradox of Organic Farming in California*. Berkeley: University of California Press.

Hamilton, M., Waddington, P. A. J., Gregory, S., and Walker, A. 1995. Eat, Drink and Be Saved: The Spiritual Significance of Alternative Diets. *Social Compass* 42: 497–511.

Howard, A. 1943. *An Agricultural Testament*. https://web.archive.org/web/20100702222720/http://pssurvival.com/PS/Agriculture/An_Agricultural_Testament_1943.pdf

Howard, S. A . 1974 (1947). *The Soil and Health: A Study of Organic Agriculture*. New York: Schocken Books.

Kareklas, I., Carlson, J. R., and Muehling, D. D. 2014. "I Eat Organic for My Benefit and Yours": Egoistic and Altruistic Considerations for Purchasing Organic Food and Their Implications for Advertising Strategists. *Journal of Advertising* 43: 18–32.

Lawler, A. 2016. *Why Did the Chicken Cross the World? The Epic Saga of the Bird that Powers Civilization*. New York: Atria Books.

Logan, W. B. 2007 [1995]. *Dirt: The Ecstatic Skin of the Earth*. New York: W. W. Norton.

Magnusson, M. K., Arvola, A., Hursti, U. K, Aberg, L., and Sjödén, P.-O. 2003. Choice of Organic Foods Is Related to Perceived Consequences for Human Health and to Environmentally Friendly Behavior. *Appetite* 40: 109–17.

Malinowski, B. 1935. *Coral Gardens and Their Magic: A Study of the Methods of Tilling the Soil and of Agricultural Rites in the Trobriand Islands*. New York: American Book Company.

McKenzie, M., Morgan, K. L., Interis, M. G., and Harri, A. 2013. Who Buys Food Directly from Producers in the Southeastern United States? *Journal of Agricultural and Applied Economics* 45: 509–18.

Northbourne, L. (Walter James). 2003 (1940). *Look to the Land*. Sophia Perennis.

Ohlson, K. 2014. *The Soil Will Save Us: How Scientists, Farmers, and Foodies Are Healing the Soil to Save the Planet*. Harlan, IA: Rodale Books.

Ohnuki-Tierney, E. 1990. The Ambivalent Self of the Contemporary Japanese. *Cultural Anthropology* 5: 197–216.

Organic Trade Association. U.S. Consumers across the Country Devour Record Amount of Organic in 2014. www.ota.com/news/press-releases/18061

Patzel, N. 2010. European Religious Cultivation of the Soil. In *Soil and Culture*, ed. E. R. Landa and C. Feller, 417–29. Dordrecht: Springer.

Paull, J. 2014. Lord Northbourne, The Man Who Invented Organic Farming, A Biography. *Journal of Organic Systems* 9.

Peters, S. 1979. *The Land in Trust: A Social History of the Organic Farming Movement*. Ph.D. Dissertation, McGill University.

Pollan, M. 2006. The Omnivore's Dilemma: A Natural History of Four Meals. New York: Penguin Press.

Riffkin, R. 2014. Forth-Five Percent of Americans Seek Out Organic Foods. www.gallup.com/poll/174524/forty-five-percent-americans-seek-organic-foods.aspx

Robbins, R. 2005. The History of Technology: The Western Tradition. In *The Oxford Encyclopedia of Science, Technology and Society*, ed. Sal Restivo. New York: Oxford University Press.

Rossbacher, L. A., and Rhodes, D. D. 2007. The Magic in Dirt. *Geotimes* 52: 60.

Schösler, H., de Boer, J., and Boersema, J. J. 2013. The Organic Food Philosophy: A Qualitative Exploration of the Practices, Values, and Beliefs of Dutch Organic Consumers within a Cultural–Historical Frame. *Journal of Agriculture and Environmental Ethics* 26: 439–60.

Seufert, V., Ramankutty, N., and Foley, J. A. 2012. Comparing the Yields of Organic and Conventional Agriculture. *Nature*: 229–32.

Shapin, S. 2006. Paradise Sold: What Are You Buying When You Buy Organic? *New Yorker*, May 15. www.newyorker.com/magazine/2006/05/15/paradise-sold

Shepherd, R., Magnusson, M., and Sjödén, P.-O. 2005. Determinants of Consumer Behavior Related to Organic Foods. *Ambio* 34: 352–9.

Shiva, V. 2008. *Soil Not Oil: Environmental Justice in an Age of Climate Crisis*. Cambridge, MA: South End Press.

Smith, T. A., Huang, C. L., and Lin, B.-H. 2009. Does Price or Income Affect Organic Choice? Analysis of U.S. Fresh Produce Users. *Journal of Agriculture and Applied Economics* 41: 731–44.

Solomon, R. C. 2003. *Spirituality for the Skeptic: The Thoughtful Love of life*. New York: Oxford University Press.

Średnicka-Tober, D., Barański, M., Seal, C. J., Sanderson, R., Benbrook, C., Steinshamn, H., Gromadzka-Ostrowska, J., Rembiałkowska, E., Skwarło-Sońta, K., Eyre, M., Cozzi, G., Larsen, M. K., Jordon, T., Niggli, U., Sakowski, T., Calder, P. C., Burdge, G. C., Sotiraki, S., Stefanakis, A., Stergiadis, S., Yolcu, H., Chatzidimitriou, E., Butler, G., Stewart, G. and Leifert, C. 2016. Higher PUFA and n-3 PUFA, Conjugated Linoleic Acid, α-Tocopherol and Iron, But Lower Iodine and Selenium Concentrations in Organic Milk: A Systematic Literature Review and Meta- and Redundancy Analyses. *British Journal of Nutrition* 115: 1043–60.

Starr, A. 2010. Local Food: A Social Movement. *Cultural Studies <=> Critical Methodologies* 10: 479–90.

Steinbeck, J. 1939 (2002). *Grapes of Wrath*. London: Penguin.

Steiner, R. 1924a. *The Agricultural Course*. Chapter 2. http://wn.rsarchive.org/Lectures/GA327/English/BDA1958/19240610p01.html

Steiner, R. 1924b. Discussion after Lecture 4. http://wn.rsarchive.org/Lectures/GA327/English/BDA1958/Ag1958_discuss4.html

Steiner, R. 2000 (1919). *Toward Social Renewal: Rethinking the Basis of Society*. Hillside, UK: Rudolf Steiner Press.

Taylor, C. 1989. *Sources of the Self: The Making of the Modern Identity*. Cambridge: Harvard University Press.

Trudgill, S. 2006. "Dirt Cheap"—Cultural Constructs of Soil: A Challenge for Education about Soils? *Journal of Geography in Higher Education* 30: 7–14.

Vogt, E. Z. 1993. *Tortillas for the Gods: A Symbolic Analysis of Zinacanteco Rituals*. Durham: Duke University Press.

Worthington, V. 2001. Nutritional Quality of Organic Versus Conventional Fruits, Vegetables, and Grains. *Journal of Alternative and Complementary Medicine* 7: 161–73.

PART FOUR

Organic Food Systems: Choice and Culture

EDITORS' INTRODUCTION

What we eat every day, and the production methods that bring our meals to the table, are part of a much larger food system that includes economics, politics, sociocultural processes, and consumer practices. Organic farming and the distribution value chains that bring food to plate are embedded within and determined by this system. But the food system is often invisible to producers and consumers; they are shaped by it but do not see its larger shape. After examining organic food and farming through the lens of history, production and values, it's time to shift perspective to ask how organics fits into the larger and complex construction of how our food reaches our plate—and why. It's time to place organics within the food system.

These chapters explore this larger system by examining the marketing of organics, adoption for institutional kitchens, and the sociological (rather than scientific) understanding of GMOs. The first chapter is written by two chefs who have actively integrated organic foods into institutional kitchens and explain the barriers and pitfalls to doing so. Given that it's easy for consumers to blandly state that their "food out" should be organic, it's refreshing to learn how getting organics on the plate isn't as easy as simply ordering from a distributor. Similarly, the marketing of organic and other forms of "ethically sourced consumables" are determined by consumer perceptions, government rules and regulations, and applicable avenues for reaching the potential buyer. All of these cultural sites and processes are constructed through social norms and practices, ensuring that the advertising of organics—and thus the messages that construct the consumer's notions of the concept of organic—is profoundly determined by sociocultural meanings and expectations. Rounding out this section is a chapter on the social understanding of GMOs. Often posed as oppositional to organics, GMOs offer an opportunity for readers to think about what is organic

FIGURE IV.1 *Produce at the farmers' market. Photo courtesy of Janet Chrzan.*

about organics, and what is conventional about GMOs and thus think more deeply about the processes that create our food. The volume concludes with an examination of the future of organic and agroecological farming. Where are we going, and how do we get there? What's possible, what's wishful thinking, and how do we distinguish between the two?

Each chapter in this section queries different facets of the food system and will hopefully raise questions for readers about how this system works, how individuals interact within it, and how our food gets to our plates.

13

Profile: Chef Steven Eckerd

This chapter demonstrates that bringing organic, sustainable, local, and ethical food into restaurants and institutions isn't easy. There are a lot of new kitchen skills that have to be learned, new ways of thinking about food, new relationships to build with farmers and distributors. The difficulties are greater for institutions because of budgets, but budgets exist for higher-end restaurants too and sometimes justifying the local and organic to a restaurant manager isn't easy. The cost of food on the plate is very carefully calculated and taking losses on specific products too many times can mean the menu gets changed by executive fiat.

A restaurant works a bit differently than an institutional kitchen but some of the same constraints are the same. Just as Chef Cohen pointed out, you still have to know your audience, do your homework, educate your staff, suppliers (and management), stay on message, be true to your vision and stay transparent, and communicate your passion for farm to table. If you can't make everyone understand why you do what you do, you'll never get a chance to do it. Two of Chef Cohen's points really resonate with me: to "be true in all stages of procurement, communication, promotion, and when educating" and "think outside of the box." Those two represent creativity and craft. They are part of expressing yourself as a chef and being clear what your product is and how to ensure that it's the same kind of food but also radically different because of the care and creativity you bring to using and serving the food to patrons. But creativity in the kitchen is part of what it means to "work with a farmer," the lost art of looking at your product and how it changes from season to season, year to year. Seasonality is so important, and our food system is set up to pretend it doesn't exist. But say you taste something amazing that really inspires you and you create a dish around it ... that one varietal may not taste like itself for four more years until conditions are correct again. Seasonality! It's a real problem within a restaurant, since products change throughout the year, so how do you give your guest the same experience consistently?

I think there is a lot of pressure on chefs to go sustainable or organic in the kitchen—that's what a lot of diners say they want—but they often don't really want to pay for it. The meal costs more when it's organic, local, and sustainable. A lot of diners will quiz the wait staff about the food before they order, they tell them that they only eat chemical-free, or organic, and they want to make sure that we're upholding their values. But how do we explain that we are sourcing from a stupendous farmer like Alex Wenger, who grows organically but isn't USDA-certified, when they ask if our food is "certified organic"? To me, the food he produces is pure and an expression of the

FIGURE 13.1 *Chef Steve Eckerd at the farmers' market. Photo courtesy of Janet Chrzan.*

land; he grows organically, but how can I explain to a diner how good his food is if the diner only wants to know if he's "certified," as if it's a check-off item on a list? So your relationship with the diner is really important; they need to know that they can trust you to source the very best for them and to consider all of the ethical and health parameters when you buy for the restaurant. I have relationships with guys who arrive at the restaurant at 8:00 a.m. with trucks full of food picked the afternoon before; it's fresh, tasty, and right of the moment. It inspires ... but it costs. That level of care and expertise in farming isn't cheap, so the food isn't cheap. So I really understand what Dr. Deutsch and Chef Cohen are writing about; it's really hard to integrate those foods into a mass-production kitchen; again, it's too expensive and people don't want to pay the premium. And the relationships that I can nurture with the small farmers won't work with the institutional kitchen, which has to be able to source larger quantities on an absolutely predictable and regular schedule. When you cook in fine dining, you can use more organic and local products because you can respond to quirky farmer product lists and charge more for the end product.

But the problem is that these foods, even for the high-end kitchen, are expensive to serve, so you must commit to regional farm to table and be able to share those values with your patrons, because the meal does end up being more expensive. Frankly, for the chef, it's easier to not use these ingredients and you can make more money by ordering from the big distributors. But I want to provide intrinsic value and uphold food traditions. It's about the morals of the chef, in the sense of giving the guest a fair product for their money based on health, taste, and ethical practices. Part

FIGURE 13.2 *Chef Steve Eckerd cooking in a home kitchen. Photo courtesy of Janet Chrzan.*

of it is making sure you are using real food, but then, since you have to provide meals to patrons, you have to understand and communicate the real value of the product after it travels down the food chain. Essentially, to stay afloat, you have to ask how to attract the 99 percent and make them need the product, since the 1 percent can already have it whenever they want.

But for both types of cooking, fine dining and institutional, you always have to "know your audience," you have to understand the cultural trends of the local population and then adapt those ideals to create your own expression of the food, the place, the terroir. To respect the dignity of the food, and its producers. To remain honest with diners, and to remain true to your vision. It's harder than you think.

FIGURE 13.3 *Chef Steve Eckerd with basket of produce. Photo courtesy of Janet Chrzan.*

Institutionalizing Organics: A Practical Consideration of Challenges and Opportunities for Sustainable Foods in Noncommercial Foodservice and Foodservice Education

Jonathan Deutsch with Budd Cohen

Organics in institutions sounds like an obvious good thing. Much of this volume, in fact, has been devoted to why sustainable foods are preferable to conventional alternatives. Who would want our must vulnerable populations—children, seniors, people with disabilities, people suffering from homelessness and/or food insecurity—all of whom rely most heavily on institutional foodservice (Tsui et al. 2014)—to be eating food we do not feel good about? Further, we do not want otherwise healthy populations like members of the armed forces, university undergraduates, or prisoners, who are required to consume the vast majority of their calories from an institutional kitchen, sometimes every meal for years, to receive anything but "good" food, however the term is defined. But operating a foodservice environment partially or wholly incorporating organics brings with it a host of challenges and practical considerations. This chapter brings the politics, policies, and polemics to the practical. Assuming we agree that we value organics and want people who rely on institutional or noncommercial foodservice (those terms are used interchangeably in this chapter), anywhere in the lifecycle, to benefit from them, how does one "pull the trigger" as a sustainable institutional feeder? And what sort of practices defines an institutional foodservice provider as a sustainable one?

When the primary author was in graduate school, he worked as a site inspector for the state emergency feeding program, working with soup kitchens, shelters, and food pantries to make sure food safety and sanitation guidelines were observed, and consulting on equipment and process improvements. At the time, the state launched a fresh produce incentive program: if an operation committed to using 15 percent of their purchasing budget on in-state produce through an institutional community-supported agriculture (CSA) program, they would receive a rebate that could be put to other uses. It was essentially an opportunity to get unspecified local produce free or with significant subsidies, so most of the agencies availed themselves of the incentive. It often worked idyllically—gorgeous apples, sweet corn, and greens making it from farm to soup kitchen, sometimes so fresh that the food was still warm from the field. At other times, though, the incentive program had cooks regretting their participation. Imagine a busy soup kitchen, getting ready to feed a few hundred people, staffed largely by good-hearted but elderly untrained volunteers with pieced-together equipment, who are faced with cases of large thick-skinned organic local rutabaga an hour before the doors open for service. In a scene that would be at home on the TV satire "Portlandia," well-meaning volunteers ended up distributing the cranium-sized tubers as-is to people who faced housing insecurity and, in most cases, could not identify the vegetable, let alone what to do with it, even if they had had kitchen access. Leaving the site and stepping over a rolling rutabaga on the sidewalk was the sustainable foods version of tumbleweed from a Western film. Good intentions, woefully misapplied.

Our goal in this chapter is to point out some common challenges and pitfalls in organics in institutional feeding in hopes of keeping turnips food and not baseballs. We will also point out traditional and nontraditional promising practices and address opportunities for educating future professionals.

Challenges

The challenges of incorporating organics—whether grown on-site, sourced through a procurement process, or sourced in combination, are many. Our goal is not to discourage the adoption of an organics program in noncommercial foodservice. To the contrary, we want to make the reader aware of common challenges so that the program can be thoughtfully and effectively deployed.

A first consideration is, of course, cost. Usually—though not always—organics cost more. There are numerous reasons for the increased cost—higher cost of organic inputs such as feed and fertilizer, more labor required (e.g., compare hand-weeding to spraying an herbicide between rows of vegetables), lower yield due to shrink from insect infestations or growing without synthetic fertilizer, the cost of the certification process (including mandated fallow periods when converting from conventional agriculture), and, when dealing with smaller producers, being unable to benefit from the economies of scale that tends to be the norm in conventional agriculture (and, to be fair, increasingly in organics as well), to name a few. To be sure, the cost differential between conventional and organic products is shrinking, but one key consideration is that those advocating for a well-established organics program, whether an institution issuing a request for proposals, consumers advocating for organic options, or diners voting with their wallets, must be prepared to bear the associated costs. Too often we hear about institutions that want their foodservice operators to move to more sustainable practices and more ethical purchasing but are not willing to put the associated resources into the program. Political priorities must match tactical support.

Another significant challenge is sourcing, especially across the seasons and with the standard nutritionist-approved cycle menus that repeat, for example, every twenty-one days throughout the

FIGURE 13.4 *Farmacology. Photo courtesy of Budd Cohen.*

year. Even in conventional purchasing, but especially in organics, not every product is available when requested. Weather, pests, transportation, fewer producers (and especially fewer certified organic producers), and other factors may cause an item normally available at a particular time of year in a region to be unavailable, in short supply, or of poor quality. Further, certified organic versions of some foods may not be available, while conventional products may be abundant and of high quality. That introduces a dilemma for a foodservice manager—keep the menu as approved, substituting a conventional product for an organic one, change a menu item from one organic product to another, or "86" (cancel) a proposed, approved and budgeted menu item. While such a decision may seem simple, it becomes challenging in institutions like hospitals, schools, and childcare centers that require a registered dietitian to approve all menus well in advance and where special diet requirements may limit the flexibility of a chef to improvise based solely on product cost, quality, and availability.

FIGURE 13.5 *Moon and stars watermelon. Photo courtesy of Budd Cohen.*

Adding local food to an organics program reveals further challenges. In temperate regions of the United States, Canada, Europe, and the geographic majority of Asia, the season for local organic produce is a very short one, from June through November, with greatly diminished selection in the winter and spring months. In other cases, local just cannot happen—for example, tropical products like coffee, bananas, and cocoa either need to be sourced abroad or removed from the menu in most settings in the United States.

As the second author—the executive chef of a sustainably minded college foodservice operation—points out in his work, organics is more work in the kitchen as well (Cohen 2014). Organic products are not as likely to be available in value-added labor saving foodservice formats such as fresh-cut (washed, trimmed, and cut to specification), individually quick-frozen, or pre-cooked. Staffing levels need to be sufficient, especially for cleaning and preparation of fresh produce. Further, staff skills and attention to detail need to be enhanced to ensure organic produce is adequately cleaned and properly prepared. For example, conventional fresh spinach typically comes to a foodservice operation triple-washed and bagged. It can go directly into a pan or salad. While organic spinach is now widely available similarly processed, if sourcing from a farm, the spinach will typically need to be cleaned in-house, a skill that may need to be retaught or reinforced as cooks gain increased distance from our food supply. We have dined on sandy spinach at even the most exclusive restaurants. With rising wages, the importance of labor costs in organic food preparation must be a central consideration.

There is also an important consumer education challenge in managing an organics program. As pointed out elsewhere in this volume, there is widespread misunderstanding regarding the nuance of the world of organics. Further, while we may know objectively that organic products may naturally contain other organic organisms such as bugs—less likely to appear in pesticide-treated produce—a guest who discovers a slug, whether dead or alive, in her or his salad may find that the guest's disgust to this common aversion far outweighs their interest in hearing the merits of organic produce from the manager on duty, further underscoring the need for a well-trained staff and extra care in preparing organic foods, as well as consumer buy-in to an organics program.

Traditional Promising Practices

Despite these challenges associated with implementing organics programs, many institutions are excelling in integrating organics and sustainable food programs more broadly. This section explores some promising practices at institutions that can serve as models for others.

Kitchen gardens and on-premise farms have become fixtures of many institutions. These facilities may be operated by foodservice staff, may be a separate unit of the organization where produce is bought by the foodservice operator through a purchase or budget transfer, or may be part of the educational or job training program of an institution like a school, university, prison, or summer camp. Some gardens or farms also serve their constituents in the form of CSAs or farm markets and/or provide food for a culinary school or job training program (Nargi 2016). There is further educational value in being able to show diners where their food was grown, what it looks like in situ, and what practices are being used to produce it. Studies have shown that children, in particular (Graham et al. 2005), and adults as well (Alaimo et al. 2008) are more likely to eat fresh produce after having been involved in its production.

Beyond growing organic produce for use in the institutional kitchen, a number of procurement options exist beyond conventional foodservice broad line distribution that can advance an institution's sustainable foods program. For example, some institutions contract with farms to provide sustainably produced foods specific to their needs. In some cases, the contracts go far beyond simple procurement agreements, with the institution helping to finance infrastructure, value-added processing or other improvements to the farm over a long-term contract (Lancaster 2016). Campus farms or long-term relationships with farms off-campus (either institution-owned or privately held) have now become commonplace. Rachel Cohen of Middlebury College compiled a list of college farms and contracts that includes, among many others: Amherst College, Augustana College, Bard College, Emory University, Hampshire College, Johns Hopkins University, Kennesaw State University, Kenyon College, Michigan State University, Chatham University, Tufts University, the University of Maryland, State University, of New York at Cortland, University of Vermont, and Stanford University as well as the Minneapolis public school system.

In the college and university foodservice world, University of Massachusetts Amherst has been frequently recognized as a leader in farm-direct procurement that has been effective not only in providing more sustainable cafeteria offerings but also more affordable options by taking advantage of long-term business relationships and seasonal bounty (Horwitz 2014). Other institutions report similar success in local procurement, ranging to as high as 36 percent spent on local farm-direct products (Algozin 2015). In addition, institutional CSA shares allow institutions to use their purchasing power to support regional agriculture. Like a home CSA, exact produce type and delivery time may vary so requires a resourceful and flexible chef. To temper the challenge of farm-to-institution sourcing through multiple agreements with multiple farms, a number of food hubs and farm aggregators have emerged to source products from multiple farms, aggregate them, and market them to institutions to deliver and invoice together, simplifying the procurement and customer service process. Many institutional feeders who pledge a percentage of sustainable or organic food procurement do so through selecting an aggregator that vets their suppliers and also provides educational materials about the producers.

Beyond Traditional Approaches

Beyond procurement, organic and sustainable foods can provide much more than nutrients in an institutional setting. In college campuses but also in private K–12 dining, assisted living, and other institutions, diners often engage with foodservice directors to make complex decisions about how best to provision the campus, balancing environmental priorities, price, palatability, and cultural relevance. In schools, the dining space becomes an extension of the classroom in these environments, teaching diners about where on the food chain they might focus their eating, how to use products with global cultural influences and applications, how to prepare healthy food at home, how a model meal might be composed in terms of portion size and ratio of macronutrients, or how to convene diners with divergent viewpoints around a table toward a common goal of an improved food system for all, to name a few (Arnett 2016).

Black and colleagues (2015) in a study of school foodservice in Vancouver, Canada, identified procurement as only one part of a sustainable foods program that also includes food gardens, composting systems, food preparation activities, food-related teaching and learning activities, and availability of healthy and environmentally sustainable food. These elements, taken together, can provide "fully integrate[d] programmes and policies that support healthy, environmentally sustainable food systems" (2379). Discouragingly, that study found that while some schools are actively engaged with elements of this formula, particularly gardening and composting, "no schools reported widespread initiatives fully supporting availability or integration of healthy or environmentally sustainable foods across campus" (2379). There is much room for growth and improvement in most institutions.

University of British Columbia's (UBC) "First Food Guide (2016)" presents a twelve-point vision for a "utopian UBC food system," quoted here at length as is applicable intact to nearly any institutional foodservice provider:

1. Food is produced in a way that upholds the integrity and health of ecosystems (including aquatic ecosystems) and does not disrupt or destroy ecosystems.

2. Animals raised for food are treated humanely and are integrated into ecologically friendly farming models.

3. Food is locally grown, produced and processed in support of local people, infrastructure and economies.

4. Food is culturally and ethnically appropriate, affordable, safe, nutritious and minimally processed.

5. Providers and educators promote awareness among consumers about cultivation, processing, ingredients and nutrition of food products in the food system.

6. Food and the food environment enhance community through opportunities for community members to interact and support one another to meet common interests and goals.

7. Food is produced in a socially responsible manner, such that providers and growers pay and receive fair prices for their products and have safe and humane working conditions.

8. There is zero waste produced by the system, in that waste is reduced to the greatest extent possible and what waste is produced is composted or recycled locally.

9. The system is emission and energy neutral.

10. On-campus food system actors work toward food sovereignty and agency, within the context of the wider food system.

11. On-campus food providers use their influence to transition the wider food system towards sustainability.

12. Students, staff, and faculty have access to opportunities to learn about the food system and to gain food production and preparation skills.

In addition to practices that improve the food on campuses, some institutional foodservice providers are supporting broader improvements to the food culture of its diners. For example, Foodtank's Algozin (2015) reports that American University's foodservice also supports community gardening and a CSA. Chefs and foodservice directors at institutional feeders are no longer just food providers—they often offer gardening and cooking classes and demonstrations to various stakeholders, lead trips to farms for foodservice staff and diners, sit on organizational sustainability councils, and serve as ambassadors who represent the institution to the public (Cohen 2014).

Educating Future Chefs, Dietitians, and Foodservice Managers in Organics/Sustainability

Until now we have focused on the foodservice opportunities for organics and sustainable food. But there is another consideration in institutional foodservice, especially in educational institutions but also in corrections and military feeders: how future cooks, chefs, dietitians, and foodservice managers are trained in organics. We have focused largely on foodservice *in* education (as well as other institutions)—let us briefly consider foodservice education.

Just as subscribing to an organics program demands a creative, flexible, and resourceful chef to use the best product available, reduce food waste, and be flexible due to seasonal and weather

FIGURE 13.6 *Farmers' market strawberries. Photo courtesy of Janet Chrzan.*

variations, sourcing challenges, and product quality variability, so to do chef-instructors and foodservice educators need to be flexible and responsive to these changes. Often culinary schools in particular fall victim to not practicing what is preached in the sustainable foods space and may showcase beautiful gardens and lecture on sustainable practices, but teach from a standard curriculum that does not prioritize sustainability (Deutsch 2016). At many culinary schools, to illustrate, learning objectives are tied to specific recipes (i.e., pie is taught through apple pie), which is anathema to using the best products of the season vis-à-vis environmental, cost, and culinary benefit. Similarly, having a garden is wonderful for a culinary training program, but is there room in the curriculum to process a few five-gallon buckets of radishes each spring (Deutsch and Billingsley 2010)?

Along the same lines, we are fortunate that nearly every standard foodservice management and culinary textbook engages with food system issues including GMOs, organics, food waste, and other aspects of sustainability. But the practices on the ground in many culinary training programs leave much to be desired. Often hampered by limited budgets and the challenges of working within the larger bureaucracy, many programs provide instruction in sustainable foodservice management but do not allow students to wholly implement what they have learned in the program's own facilities. O'Donnell et al. (2015) suggest that a Food System Sensitive Model needs to be employed that can question dominant paradigms on food waste and find usable outlets for co-streams, by-products of the primary food production. For example, it is wonderful to teach culinary students to cut perfectly diced potatoes but equally important to teach what might be done with the odd-shaped trim. Put in the context of a culinary or foodservice training school, rather than starting with the recipe and ordering the associated food to prepare to meet the intended learning outcome, available food already in the system should be used first and the instructor should align that product to stated learning objectives if possible.

Student Movements/Activism

Unfortunately, whether out of good intentions poorly executed (as in our rutabaga example that opened this chapter), or shoddy business practices, there are numerous examples of organics programs falling short. Many foodservice vendors are accused of "greenwashing," performing practices that *appear* to be sustainable but actually are not. A typical example would be asking diners to separate garbage in the dining room into materials bound for landfill, plate waste for post-consumer composting, and recycling (plastic, metal, and paper), to suggest an efficient and participatory disposal system, where in reality the separate streams are later comingled and discarded together because the facility lacks a proper waste management program. Similarly, materials like compostable takeout containers are only green if a working post-consumer composting digester exists. Otherwise they are just "greenwashed" take-out containers that will be landfilled along with their Styrofoam cousins. In other cases, greenwashing is not merely *suggestive* of practices being more sustainable than they are, but downright fraudulent, as in conventional offerings being improperly labeled as organic, or "local," used freely (though it is true that everything is local *somewhere*), or conventional products substituted over time. A real-world example of the last would be a fair trade coffee program with branded urns indicated both the brand and fair trade status of the coffee. A new manager arrived and deemed the coffee program too expensive, switching to a more cost-effective alternative but not replacing the previous brand's equipment, complete with Fair Trade designation.

Institutional diners—particularly college students—are increasingly (and rightfully) skeptical of shady sustainability practices and have been vocally objecting to them on many campuses (Fertig 2015). Concerns about sustainability and ethical sourcing often intersect with concerns about foodservice workers' wages, benefits, and conditions (O'Neil 2016); culturally relevant menus (Friedersdorf 2015); and food safety and sanitation (Johnson 2015). These concerns have prompted support for worker strikes, protests, editorials, and, more constructively, diner representation on foodservice contract and oversight committees; grassroots projects such as the food recovery network, where students recover surplus food from dining halls and distribute it to people in need (Food Recovery Network 2016); and support for foodservice workers, particularly from graduate students, in the form of legal and social services.

Conclusion

Incorporating organics in institutional feeding has challenges, to be sure, but can also advance the culinary, sustainability, and financial goals of the institution while providing high diner satisfaction.

The Institute for Agriculture and Trade Policy produced a sustainable farm-to-hospital toolkit, which advised strategies for mitigating the higher cost of organics in the institution's budget. Among them: take advantage of seasonal bounty where seasonal local produce may actually be cheaper; reduce spending on health-destructive products such as frying oil or unsustainable products such as paper tray liners; reduce food waste; streamline inventory; serve more plant-based proteins; and adjust retail and catering pricing (Kulick 2013). These suggestions are not given to suggest that a foodservice operator will have an easy road to sustainability; just that it is being done, and done well in many settings.

In our experience, achieving proper balance is important. Sustainable foodservice is a goal to which we must constantly and vigilantly strive. There is no magic moment when full sustainability is achieved—practices can and must always be improved. Similarly, while there is never "good enough" in sustainable foodservice, there is constant compromise and "better than." Diners have a host of considerations they bring to a meal—environmental impact, for sure, but also budget, flavor, nutrition, multi-sensory appeal, social function, and desire. Sustainability alone cannot guide a dining program. As an example, there is no question that from an environmental standpoint, a glass of tap water is the most sustainable beverage option for campus dining—low cost, no empty calories, no packaging, and so on. Certainly every sustainably minded foodservice operation should have free filtered tap water available for its guests. But unless the reader never savors a beer, cup of coffee or tea, kombucha, or glass of juice, you will understand the importance of compromise. Having filtered tap water available for free is an improvement over making only bottled water available for a fee; sourcing fair trade organic coffee is preferable to unethically produced coffee; but eliminating coffee from a college campus due to the greenhouse gas emissions produced in its production and transportation would be a step many would find too extreme. So education and compromise are key.

FIGURE 13.7 *Tomato. Photo courtesy of Budd Cohen.*

It is important for foodservice operators to be forthcoming regarding why they make the sourcing and operations decisions they do, where they compromise and why. It is important for them to share materials from the manufacturers and distributors who supply them regarding how their products are produced and distributed. And it is the duty of the diner to remain engaged, vigilant, constructive, and participatory to collaboratively improve the practices of the operation.

Chef Budd Cohen's Top Ten Guide to Running a Farm-to-Institution Operation

1. Know Your Audience—Communicate directly (in person) and through electronic means with the end users of your product. Getting feedback is important. Gathering demographic data will also help guide your purchases of goods, and your menu offerings. A very broad example would be: children consuming food at a preschool or kindergarten have different preferences than adults who eat at a Senior Living Facility. In either case there are many local foods that could be incorporated in their diets. You will be successful when you meet them and understand their needs.

2. Do Your Homework—the Internet is a great resource for information about current events regarding all things farm-to-table. Searches can be easily customized for local or global information. Websites, including USDA and state agriculture departments, large and small farm operations, distributors, and news based can be very informative. There is practically endless information about farm and trade politics, agricultural methods, nutrition, seed preservation, organic production, GMOs, sustainable seafood, cooking, and the business side of the food and beverage industry. Live interaction is a key element as well. Anywhere you live there is some resource available. Examples may be visiting a farm market in a big city or traveling to a local farm.

3. Educate your foodservice group, vendors, and your consumers on all levels. Bring in new products and share them with your staff. Taste them, talk about their origin, their nutrient value, cost, the best ways to serve, flavor profiles, and their seasonality. Share this information with your superiors, clients, and guests. Put out samples, displays, and flyers that educate and talk to whoever is willing to listen. Let your vendors know what you are doing and they may be able to help you with more products.

4. Communicate Your Passion and share your enthusiasm for farm to table. You and your team are the most important tool for marketing and promoting your farm-to-table program. If you can motivate your staff and give them the products and tools to implement your plans they can help position your program for achievement.

5. Accept Your Successes and Failures—It takes time to find your groove. Roadblocks will come your way. Using farm-direct products in your institution will cause your cooks to work harder compared to using packed, possessed goods that they may be used to. Your supervisor or client may expect that farm-direct products will cost more and they may

not understand the value. Your end users may also feel the same way. There will be folks that support and champion the cause as well.

6. Stick with the Basics—Apples, honey, eggs, tomatoes, and cucumbers are great items to start. They are identifiable and acceptable to most diners. These ingredients are also easy for your chefs and cooks to incorporate.

7. Stay on Message—Sustainably, locally produced goods keep money in the local economy, conserve fossil fuel, keep small family farms in business, and preserve open space. In addition, when we eat food that is cooked from scratch we can monitor how much fat, sugar and salt goes into it. Promoting health and nutrition is always a good thing. It is the food, the farms and the farmers that are the star here, not you.

8. Be True in all stages of procurement, communication, promotion, and when educating— Strive for transparency with regards to where products were produced/grown, what growing methods were used, how you purchased it, and how you prepared it. Set up a system based on truth, when you list a specific farm, or method on your menu make sure your product matches your claims. Make amendments to your menu, even during the daily service, if need be. Once you have established a pattern and standard, it becomes easy to be honest, and your staff will follow suit.

9. Think Outside of the Box—Bison Heart and Stinging Nettles are two ingredients I have used recently in my cafeteria to open up conversations with my guests and staff. Use ingredients that are cost effective, local, and sustainably produced. Be daring and thrifty, stretch typically expensive proteins into stews, soups, or casseroles. Buy in-season items or #2 grade products when viable, such as peppers or tomatoes. I find these two vegetables work great in soups, salads, entrees, condiments on pizza, sandwiches, and pasta dishes.

10. Keep Health and Nutrition in Mind—The western style diet we consume, high in fat, sugar, and salt, has contributed negatively to our health. Diabetes, high blood pressure, high cholesterol, and cancer, in many cases related to obesity, are at epidemic proportions in America. Processed foods that have proliferated the supermarket shelves are generally high in fat sugar and salt. Cooking from scratch with farm-to-table products ensures that we know what is going into our food and we can choose what that is. We need to eat more whole grains, more plants in general, leaner meats and real food.

Acknowledgment

Thanks to Otitoyeni Olatunji and Ben Fulton for their valuable research assistance.

References

Alaimo, K., Packnett, E., Miles, R. A., and Kruger, D. J. 2008. Fruit and Vegetable Intake among Urban Community Gardeners. *Journal of Nutrition Education and Behavior* 40: 94–101.

Algozin, C. 2015. 12 College Campuses Leading the Way for Sustainable Dining. *Foodtank*. http://foodtank.com/news/2015/08/twelve-college-campuses-leading-the-way-for-sustainable-dining (Accessed November 18, 2016).

Arnett, L. 2016. What Gen Z Wants from Foodservice: Gen Z Is Having It Their Way, All Day—And Night—at Colgate University. *Foodservice Director*. www.foodservicedirector.com/ideas-innovation/emerging-trends/articles/what-gen-z-wants-from-foodservice (accessed November 18, 2016).

Black, J. L., Velazquez, C. E., Ahmadi, N., Chapman, G. E., Carten, S., Edward, J., Shulhan, S., Stephens, T., and Rojas, A. 2015. Sustainability and Public Health Nutrition at School: Assessing the Integration of Healthy and Environmentally Sustainable Food Initiatives in Vancouver Schools. *Public Health Nutrition 18*: 2379–91.

Cohen, B. 2014. *Farm to Table for Schools*. Scotts Valley, CA: CreateSpace.

Cohen, R. 2016. Responses to COMFOOD-L Inquiry about College That Contract with Local Farmers to Grow for Them. *COMFOOD-L*.

Deutsch, J. 2016. Revolutionizing Culinary Education: Can Cooking Save Our Food System? Proceedings of the 3rd Dublin Gastronomy Symposium, Dublin Institute of Technology.

Deutsch, J. and Billingsley, S. 2010. *Culinary Improvisation: Skill Building Beyond the Mystery Basket Exercise*. New York: Pearson.

Fertig, J. 2015. This Maine Co-op's Trying to Bring Fresh, Local Food to 10,000 College Students. *Yes! Magazine*. www.yesmagazine.org/new-economy/this-maine-co-ops-trying-to-bring-fresh-local-food-to-10000-college-students-20151109 (accessed November 18, 2016).

Food Recovery Network. www.foodrecoverynetwork.org/ (accessed November 18, 2016).

Friedersdorf, C. 2015. A Food Fight at Oberlin College. *The Atlantic*. www.theatlantic.com/politics/archive/2015/12/the-food-fight-at-oberlin-college/421401/ (accessed November 18, 2016).

Graham, H., and Zidenberg-Cherr, S. 2005. California Teachers Perceive School Gardens as an Effective Nutritional Tool to Promote Healthful Eating Habits. *Journal of the American Dietetic Association* 105: 1797–800.

Horwitz, S. 2014. Affordable, Farm-Direct Local Foods Sourcing. *Farm to Institution New England*. www.farmtoinstitution.org/blog/case-study-umass-amherst (accessed November 18, 2016).

Johnson, D. 2015. Hans Inspection Reveals Foodborne Illness Risks. *The Triangle*. http://thetriangle.org/news/hans-inspection-reveals-food-borne-illness-risks/ (accessed November 18, 2016).

Kulick, M. 2013. *Financial Strategies for Incorporating Sustainable Food into a Hospital's Budget. Sustainable Farm-to-Hospital Toolkit*. Minneapolis, MN: Institute for Agriculture and Trade Policy.

Lancaster, A. 2016. Farm-to-Table and Beyond: Operators are Teaming Up with Farms for More than Just Food. *Foodservice Director* 32.

Nargi, L. 2016. A Striking Number of College Students Are Food Insecure. Can Campus Farms Help? *Civil Eats*. http://civileats.com/2016/06/23/a-striking-number-of-college-students-are-food-insecure-can-campus-farms-help/ (accessed November 18, 2016).

O'Donnell, T., Deutsch, J., Yungmann, C., Zeitz, A., and Katz, S. 2015. New Sustainable Market Opportunities for Surplus Food: A Food System-Sensitive Methodology (FSSM). *Food and Nutrition Sciences* 6: 883–92.

O'Neil, L. 2016. Harvard's Striking Workers Have a Secret Weapon: Students. *Slate*. www.slate.com/articles/business/dispatches/2016/10/harvard_s_striking_workers_have_a_secret_weapon_students.html (accessed November 18, 2016).

Tsui, E., Wurwarg, J., Poppendieck, J., Deutsch, J., and Freudenberg, N. 2014. Institutional Food as a Lever for Improving Health in Cities: The Case of New York City. *Public Health* 129: 303–9.

University of British Columbia. UBC Sustainable Campus Food Guide. https://sustain.ubc.ca/sites/sustain.ubc.ca/files/images/UBCSustainableCampusFoodGuide.pdf (accessed November 18, 2016).

14

Profile: Alex Wenger, Organic Farmer

❝ 'Fiori d'inverno,' or 'winter flowers.' Radicchio from Treviso, grown in #lancasterpa." Within fifteen minutes after posting a photo of ruby-red candy-cane-like Trardivo and a pale golden and pink-speckled Castelfranco radicchio rose on Instagram, my phone started blowing up. Three different chefs wanted to know if I could deliver some of these tasty, ornamental, bitter, winter greens later in the week.

Marketing is an essential skill in farming. It is especially important for small farms who do not work with a buying co-op or larger wholesaler. Farms tend to grow commodity crops, where they compete at lower price points relying on high-volume sales. Or, farmers can opt to market niche products that can be sold for higher prices, but that are often more labor-intensive. Every week I send out a produce availability list to my customers, make calls and send texts to tell them about our colorful heirloom varieties and flavorful "new" crops. These are harvested to their tastes, packed and loaded in my truck, and delivered individually to each chef and restaurant.

I will never forget the first year that we grew hundreds of pounds of Quaker Pie Squash, a near-extinct historic local variety. We needed seeds from these fruit for the Roughwood Seed Collection to sell to the public. This rare variety went extinct in Pennsylvania, only to be rediscovered as a variety that had been preserved by Quaker missionaries in Kenya. Each rare squash needed to be split in half with a machete or knife and the seeds scooped out and dried.

For this project, we enlisted the help of some loyal friends, along with Chef Steve Eckerd, so that he could roast the squash halves in his restaurant to offer a vegetable world-exclusive. Planting and harvesting hundreds of milk-carton-sized squash by hand, loading them into our truck, and driving to Philadelphia, along with the splitting and seed cleaning was incredibly labor-intensive. It would have been faster to throw the squash in a produce bin, drop it off at the produce auction, and wait for a check in the mail. The trade-off was a much bigger paycheck, as we sold the squash for several times what their commodity cousins could have been sold for, as well as the satisfaction as being the only place in the continental United States where people could find this variety to eat.

We have had some great successes like the Quaker Pie Squash, or our Floriani corn. Nevertheless, even with all of the unique varieties that we grow, and our novel marketing strategies, selling fresh vegetables is fast-paced, complicated, and competitive. Market prices often change from week to week. Huge volumes of cheap, local veggies often flood the market unpredictably, driving down prices if your customers decide to cut a deal for some cheap produce.

FIGURE 14.1 *Alex Wenger at the Fields Edge Farm. Photo courtesy of Janet Chrzan.*

The fact that niche crops are also low-volume creates a challenge, and I have been left with cases of beautiful fresh-picked vegetables that needed to be sold or preserved, while I was completely sold out of the exact same crop a few days before. Those days I end up racing around trying to "find a home" for all of it so it doesn't spoil. There is nothing more soul-crushing than watching your hard work, blood, sweat, and tears rot in the compost pile. In this crazy, ever-changing market I rely on social media, my cell phone, and texts to stay on top of everything. Instagram is a useful marketing tool, but lately, with more farmers entering local niche markets, I have to tell my vegetables' stories in a different way. The problem was that other farms and regional distributors would see the photos and stories that we post on Instagram, and directly copy the crops that we grow. Then they would offer these crops in the same market, which began a race on the agricultural treadmill.

To stand apart, the name of the marketing game is education. Even though I follow the guidelines for USDA-certified organic production, I do not pursue certification because I have face-to-face interaction with my customers. I can tell them how and why we choose to grow with no synthetic chemicals. I can walk them through our fields and let them taste the results. I have built trust, and a brand based on the quality of what I grow, by being reliable and good to my word. Even though I use modern communications to share my message, this type of marketing is "old-fashioned" in many ways. I prefer it to having organic certification, and sending vegetables off to people whom I will never meet.

FIGURE 14.2 *Squash in fields. Photo courtesy of Janet Chrzan.*

I love food. Every ounce of my soul wants to create new and unexpected combinations. My loyal customers give me the opportunity to do this, when we collaborate as I introduce them to a new vegetable variety, a new crop species, or ideas for new cooking techniques based on observations from the field.

Our most creative marketing and farming experience thus far was a series of farm-to-table dinners with Chef Steve, where we grew the most special, rare vegetables we could find, then offered a curated experience to diners so that they could learn about the history of their food

FIGURE 14.3 *Alex Wenger tasting a vegetable. Photo courtesy of Janet Chrzan.*

and why we farm sustainably. Five courses ranged from a greens salad of forced winter roots, to "hairy," over-wintered, sprouted potatoes. The series was a success all around, thanks to this close collaboration between Chef Steve and our farm. We short-circuited the commodity chain by working together and creating an amazing experience for all those involved.

Make no mistake: farming is a business in the twenty-first century. It relies on efficiency, but also creativity. To do something different than sell into an industrial marketing chain is hard. Niche marketing is essential for all farmers who are ready to break the mold, and make a living from farming once again.

Marketing Organic Food in the United States

Christina Callicott and Catherine M. Tucker

Introduction

Why do US consumers buy organic food? The answer may seem self-evident to devoted consumers. Yet as organic markets have expanded, marketing research has found that multiple factors shape consumer decisions. As indicated by other chapters in this book, US consumers who purchase organic foods are often motivated by values and concerns encompassing health and nutrition, environmental sustainability, social justice, and food safety. With few exceptions, marketing through the mass media—such as TV or radio ads—has not been a major factor promoting organic food sales. Instead, countercultural and alternative food system movements interacted with consumer concerns to make organics a major growth segment in food markets by the latter part of the twentieth century, largely through word of mouth. More recently, marketers of organic foods have used packaging and presentation to attract new customers, and marketing firms are becoming increasingly interested in how to market organic foods to a growing customer base (Kareklas et al. 2014). To understand organic food markets therefore requires consideration of the evolving perceptions and critiques of the dominant US food system among consumers and farmers in the face of increasing corporate domination of markets. The following discussion will consider how the history of food-related social movements shaped the emergence and ongoing evolution of organic markets, and how government certification has affected organic markets and spurred current processes of corporatization and commodification of organic goods. Given that marketing strategies require understanding the consumer base, the discussion explores the characteristics of organic food consumers, and how values and lifestyle choices resonate with purchasing patterns. Finally, the chapter considers the ways that marketing aims to expand organic consumption by appealing to consumer values, and the current processes that have created concerns about marketing of organics as well as for the future of organic agriculture.

Historical Precedents and Developments of the Organic Good Sector in the US Markets

Modern organic food markets have emerged from alternative food movements in the United States going back at least two centuries. These movements have been motivated by concerns for political and social justice, human and animal health and welfare, and environmental stewardship, through which the production, purchase, and consumption of food became tools for the expression of moral values. Early forms of morally driven consumerism included abolitionists' boycotts of "slavery sugar" during the eighteenth century (Wilk 2010). Starting in the late 1800s, the Food Reform Movement rejected the increasing industrialization and urbanization of America, advocating a back-to-the-land ethos, protection of nature, healthy eating, and even urban gardening. It promoted a vegetarian or low-meat diet believed to better reflect the diets of early humans (Vogt 2007)—a precursor to the recent "paleo diet."

In the early twentieth century, the Dust Bowl forced American agronomists and land managers to acknowledge the connection between sustainable farming and landscape ecology. The rising dominance of industrial agriculture after World War II, with its reliance on synthetic pesticides derived from military technologies, was one of the major factors driving consumer interest in organic agriculture. The early organic movement emerged in the 1940s, as Jerome Rodale founded the magazine *Organic Gardening and Farming* and promoted the adoption of "non-chemical" agriculture (Sustainable Agriculture Research and Education 2012).

The contemporary organics movement in the United States took its modern form beginning with the 1960s counterculture, as radicalized youth rejected industrialization and commercialization, emphasized environmental values and social justice, and discovered that dietary change could have political, economic, and social consequences (Belasco 2007; Guthman 2008, 2014; Lockeretz 2007). In a world where the personal was political, decisions about sources and types of food, as well as forms of food preparation, became tools for fomenting change. With their appreciation of hedonic pleasures, the counterculture seized on the idea that revolutionary action could taste good too. They embraced ethnic foods and culinary creativity, creating what food scholar Warren Belasco describes as a "countercuisine" (Belasco 2007). Interest in sustainable, alternative agriculture contributed to a resurgence in farmers' markets during the 1970s, as consumers and activists sought locally grown food and ways to support small-scale farmers (Robinson and Hartenfeld 2007; Alkon and Norgaard 2009).

Food was also linked with racial justice through the Black Panthers' free breakfast program (Alkon 2007) and the United Farm Workers' successful boycotts of California table grapes and lettuce, in which they protested the use of toxic pesticides and the dangers they posed to migrant farmworkers (Sligh and Cierpka 2007). These campaigns helped establish social justice as a core value of the early organic foods movement.

The Development of the Contemporary Organics Market: The Countercuisine Goes Mainstream

During the last three decades of the twentieth century, interest and involvement in organic agriculture grew concurrently with increasing concerns about toxic pesticides and waning agricultural productivity. Much of the growth took place in California, a center for countercultural activity as well as agriculture. The farm crisis of the 1980s, along with commercial development pressure on farmland, drove farmers to maximize their crop value per acre. Organics, with their price premium, helped do that (Guthman 2003). In 1989, public outcry over the risks of Alar, a pesticide commonly used on apples, fueled interest in organic foods and helped quadruple California's organic acreage in two years (Guthman 2003).

The early organics market grew out of a close relationship between small farmers and their purchasers, whether they were restaurants, small health-food specialty stores, or consumers themselves (Aschemann et al. 2007). However, as the market grew, the increasing number of producers and increasing distance between the farm and the consumer led to the need for a system that would ensure the integrity of the claims that growers made about their foods. At first, various states and producer collectives developed their own systems of certification. This created a patchwork of conflicting standards and certifications, that, along with unresolved issues around processed foods and animal products, limited growth of the market (DiMatteo and Gershuny 2007; Guthman 2014). In order to better facilitate interstate and even international commerce, the federal government in 1990 passed legislation (the Organic Foods Production Act) providing for the formation of a nationwide organics certification label that would replace state labeling programs (Greene 2007).

The growing demand for organic produce also facilitated the entry of large-scale producers who could take advantage of economies of scale (Aschemann et al. 2007). Unlike the early movement that was built by small-scale farmers, many of whom subscribed to a countercultural anti-corporate ideology, these large-scale producers relied on many of the same production, processing, and distribution techniques that characterized mainstream agriculture (Buck et al. 1997; Guthman 2003, 2004).

USDA Certification and the Conventionalization of Organics

The USDA organic label was envisioned as a marketing tool to support the price premiums that organic production required and to facilitate the growth of the market beyond its original niche into the mainstream (Guthman 2014). As such, it obscured some of the more radical goals of the early organics movement, which rejected and sought to replace the industrial model of agriculture with a smaller-scale, holistic system of agroecology. Other goals such as farmworker justice also fell by the wayside (Guthman 2014). Although the USDA organics definition did not ban or outlaw these sorts of efforts, it created a system in which economies of scale could squeeze smaller producers out of the market. Jaffee and Howard (2010) describe this process as the "corporate co-optation of organic."

The development and launch of the federal organic standards stimulated a cascade of changes in the organics market that increased access to organic foods for a wider spectrum of consumers. Changes were already taking place due to market growth and maturation, and corporate actors had begun eyeing the organic food sector as its growth through the 1990s outpaced other sectors of the food industry (Guthman 2014). The release of the first draft of the organics standards in 1997 initiated a series of corporate acquisitions of organic food companies as well as mergers and strategic alliances (Buck et al. 1997). The consolidation of the organics market, and particularly the

FIGURE 14.4 *USDA organics label.*

increasing participation of large corporate actors, has been obscured by what Howard (2009) calls "stealth ownership," whereby the corporate owners of organic brands do not reveal themselves in the labels, packaging, or promotional materials associated with their products.

In 2001, the long-awaited phase-in of the USDA label set off a process of concentric diversification in the organic food industry, in which mass-market brands introduced organic versions of their top products, or created private-label lines of organic foods (Howard 2009). These included organic versions of mainstream staples such as Nabisco Oreos and Kraft Macaroni and Cheese. This process accelerated when Wal-Mart announced its entry into the market, promising to retail organic products for no more than 10 percent above the price of conventional brands. As of this writing, mass-market outlets such as Costco and Wal-Mart claim the majority of retail organic sales, offering lower prices and greater accessibility to a wider range of consumers. Increasing availability of organic products through corporate food purveyors has been recognized as a democratizing process (Martin 2014) that has expanded demand. The ongoing market trends of commodification, industrialization and mass marketing has come to be known as "conventionalization of organics" (Guptill 2009; Howard 2009). The process includes a weakening of organic markets' traditional ties to anti-corporate social movements, dilution of consumers' desires for alternatives to conventionally produced foods, and increased opportunities for capital accumulation by off-farm entities (Howard 2009).

Who Buys Organic? Consumer Characteristics and Values in Expanding Markets

Organic food sales have been growing by double digits, reaching $43.3 billion in the United States in 2015 (Organic Trade Association 2016). Today, US consumers can purchase organic food at more than 20,000 natural food stores, and three of four conventional grocery stores carry some organic choices (USDA 2016a). The increasing availability of organic foods and growing consumption has spurred numerous studies endeavoring to define the demographic profile of the "organic food consumer" around the world. Comparatively few studies of organic food consumers have taken place within the United States, and they reveal high heterogeneity in US organic consumerism. Despite assumptions that wealthy, well-educated, and white consumers purchase more organic products than other groups, surveys show that organic consumption occurs across a range of ethnic, educational and income groups (Lockie et al. 2006). A Bellows et al. (2008) survey of randomly selected households found that older, lower-income and less educated respondents are among the regular purchasers of organic foods, and that respondents who identify as Asian/Pacific Islander, Native American and "other" bought organics more frequently than did other ethnic groups. Dimitri and Dettmann (2012) found that higher income is positively correlated with organic consumption; however, they also found that half of all organics consumers have an income below $50,000. While very low income may be a barrier to frequent organic consumption, organic consumption appears to decline at very high income levels (Mintel 2015). Education appears to be a fair predictor of organic consumption (Dimitri and Oberholtzer 2009; Curl et al. 2013), yet Mintel (2015) reported that consumption of organic food declined at the highest education levels. Perhaps the best predictor of organic consumption is age; almost all studies find that younger people tend to buy more organic foods than their older counterparts do. Proximity also matters; two studies found that those who live within five miles of a Whole Foods store buy organic foods more often and purchase a wider variety than those living further away (Dimitri and Dettmann 2012; Curl et al. 2013).

The above-cited studies as well as a number of others seek to determine what values drive a person's decision to purchase and consume organic foods. Along these lines, the findings are fairly consistent: Most people buy organic foods because they think that they are healthier than non-organic foods (Dimitri and Lohr 2007; Van Loo et al. 2012; Aschemann-Witzel et al. 2013; Rodman et al. 2014; Mintel 2015). The notion of "health" may refer to both enhanced nutrition content as well as the absence of toxic synthetic compounds, and studies do not consistently differentiate between the two in measuring consumer priorities. Other concerns include environmental sustainability, animal welfare, food safety, food quality, taste, worker welfare, and tradition (Lockie et al. 2006; Pearson et al. 2011; Hartman Group 2014; Mintel 2015).

The Attitude–Behavior Gap and Barriers to Organic Consumption

Although studies find that most people cite health as their main reason for buying organic, their findings may not be an accurate reflection of the broader interest in organic foods. Bellows et al. (2008) found that 73 percent of their survey respondents who highly value organic production methods do not buy organics regularly. This finding points to a phenomenon known as the attitude–behavior gap (Bellotti and Panzone 2016), which vexes sociologists studying a range of issues from organic food purchases to pro-environment attitudes and behavior. In Bellows et al. (2008), those who valued organics but did not buy them included the religiously observant, as well as older, lower-income, and less educated respondents. The authors suggest that certain individuals value the public benefits (environment, social, and animal welfare) over the private benefits (health) of organic foods, and that because they are not regular consumers, they are "alienated politically and from market analysis" (Bellows et al. 2008: 19–20).

Lifestyle Choices, Consumer Values, and the Appeal of Organics

Efforts to establish values associated with organic food consumption have thus identified links to a wide-ranging set of concerns, which seem to defy assumptions of a typical organic food consumer. In this context, some researchers have taken a more integrative approach, seeking to understand the interactions of ethical values, norms, beliefs, and attitudes that shape food choices (Honkanen et al. 2006). Along these lines, several studies have characterized the values expressed by many consumers of organic foods as an expression of "universalism" (favoring the well-being of all people and nature) that integrates altruism (concern for others), ecology (as interest in harmony with nature and a sustainable future), benevolence toward others, and spirituality (Makatouni 2002; Zanoli and Naspetti 2002; Fotopolous, Kyrstallis and Ness 2003). While much of this information emerges from studies in the United Kingdom, similar values appear to resonate with US consumers. This conjunction of values is associated with personality measures and attitudes that inform lifestyle decisions and behaviors, which extend beyond organic food consumption (Hughner et al. 2007). Thus consumers who embrace core values including health, environmental sustainability, and social justice appear more likely to recycle (Lockie et al. 2004), and may also adopt vegetarian diets and other environmentally conscientious behaviors (Honkanen et al. 2006). More broadly, committed consumers of organic foods appear to be embracing an alternative lifestyle that could be interpreted as a critique of the dominant, individualist-oriented value system and mainstream consumption patterns.

Expressions of alternative lifestyles associated with organic food consumption range across a continuum of countercultural groups to more mainstream consumers motivated by environmental, health, and social justice concerns. Countercultures with a distinctive food ideology include punk culture, which favors a cuisine that includes organic and locally grown produce along with acquisition of pilfered and discarded foods. This combination of food sources jointly points to punk culture's rejection of conventional norms and power relationships (Clark 2004). For some mainstream consumers, decisions to buy organic food link to interest in Fair Trade and other nonconventional trade arrangements. These consumers often seek alternatives to conventional markets as a way to support human rights and social justice for struggling farmers and agricultural workers (Tucker 2011; Moberg and Lyon 2013). Given consumer interest in organics, certain Fairtrade-certified goods such as coffee and cocoa are now likely to be produced and sold with organic certification as well (Fairtrade International 2016). Other mainstream organic consumers have been supporting farmers' markets as a venue for purchasing organic and locally produced foods and other goods. Farmers' markets have been growing across the United States; between 2004 and 2016, the number increased from about 3,700 to 8,669 (Robinson and Hartenfeld 2007; USDA 2016b). Consumers interested in sustainable local food systems may patronize farmers' markets for reasons that go beyond interest in health or pesticide-free food, such as community building (Robinson and Hartenfeld 2007). In short, underlying values appear to be more important for understanding organic consumerism than ethnicity, socioeconomic status, education, age, or other demographic variables (Atkinson and Kim 2014).

Marketing and Promotion Strategies in the Organic Food Industry

Marketing strategies for selling organic foods pay close attention to consumers' expressed values. Media-based advertising has yet to become a major strategy for the promotion of organic foods (Lockie et al. 2006; Van Loo et al. 2012). However, the industry uses a number of other direct and indirect strategies to target consumers and appeal to values that have been closely associated with the organic movement: health, environment, and animal welfare, as well as small-scale farming and a decentralized food system. Some marketing strategies may represent a concerted attempt to meet consumer demand, while others obscure the true nature of the industry, leading to charges of greenwashing that threaten the credibility of the organics label and consumer faith in the industry as a whole (Mensing 2008; Schuldt and Hannahan 2013). As corporations now dominate organic markets, mass-media advertising appears likely to play a greater role, and advertisers are working to find marketing strategies that appeal to consumers (Kareklas et al. 2014). For example, one study found that egoistic (personal health) and altruistic (pro-environment) interests were both important variables in predicting consumer interest in organic foods. When exposed to a range of sample ads highlighting different messages, including a neutral control ad, consumers perceived ads featuring both altruistic (pro-environment) and egoistic (personal health) messages more favorably than ads featuring solely egoistic (personal health) claims (Kareklas et al. 2014).

Currently, the marketing and promotion of organic foods comprises a number of diverse strategies aimed at building brand awareness and making organics more accessible. Packaging is one of the most important conveyors of brand image and narrative, but it can also be the site of what critics consider to be false advertising. Perhaps the most important element of organic packaging is the USDA organic label itself, which acts as a sort of "meta-brand" and is highly influential in consumer decision making (Van Loo et al. 2011, 2015; Schleenbecker and Hamm 2013). However, consumers may associate the USDA label with certain qualities, such as superior nutrition and food safety, that it is not designed to represent (Guion and Stanton 2012). Indeed, critics suggest that this misunderstanding may represent a form of de facto false advertising that misleads consumers and undermines attempts to create food production and labeling systems that could more accurately reflect consumer expectations (Friedland 2005). Other critics claim that the label is underwritten by an inadequate system of certification, particularly in the international market. Supplier nations often lack the political, economic and infrastructural conditions necessary to meet regulatory requirements, greatly increasing the risk of fraud (Liu 2011).

In attempting to meet consumer demand for a lower price point, the industry has introduced lines of nonorganic "natural" foods, which compete strongly with certified organic foods for consumer dollars as well as market credibility. Abrams et al. (2010) found that consumers did not understand and mistrusted the "all natural" label but were nevertheless more likely to purchase those products than organic ones. The authors of this study suggest that "all natural" is used to add false value to conventionally produced products and to capitalize on the association between nature and organic foods (Abrams et al. 2010). Consumers may be complicit in the misreading, as they use the "veneer of green" to help them rationalize and justify purchases that they might otherwise avoid (Atkinson and Kim 2014). The issue has material repercussions for the organics industry, as market growth for "natural" products outperforms that for organics (Mintel 2011).

The branding of organic foods through packaging as well as through web-based promotions often builds on a narrative of small farms, family-owned business and community engagement. One example is the previously mentioned "stealth ownership" (Howard 2009) in which corporate actors choose to obscure their acquisition and ownership of companies that first established themselves as family-based and cooperative ventures. Such acquisitions include Horizon Organic and Earthbound Farms, bought out by WhiteWave Foods (a spin-off of Dean Foods) (Howard 2016). Another example is the use of trompe l'oeil in packaging designs, visual effects that create a sense of animation, interaction, and engagement between the consumer and the product (Wagner 2015). For example, think of a frozen pizza container showing a whole pizza fresh from the oven, perhaps with a slice half pulled away as melted cheese drips from its edges. The marketing firm Hartman Group (2010) lauds Annie's Homegrown Mac and Cheese (a General Mills product) for branding themselves with a sense of authenticity, a founding narrative, transparency, playfulness, and a history of active engagement with their customers. Similarly, Johnston et al. (2009) found that marketing of corporate organics "consistently draws on food democracy images and narratives, connecting products to a particular locale and family farms, and highlighting a 'personal' history behind the brand while obscuring spatially dispersed commodity chains and centralized ownership structures" (p. 512). Finally, companies use their websites and the internet to build relationships with and among their consumers. The internet is a common source of health information for the public; therefore, it is also a good place to press health-related claims regarding food (Mintel 2010).

Supply Shortages, Increasing Imports, and Challenges to the Viability of the Organic Brand

While organic products still constitute only a small percentage of the overall food and beverage market, the rapid pace of growth over the last two decades has led to several challenges to the credibility of the label. The increasing demand for organics and the slow conversion of farmland to organic production has led to shortages of supply, including the grains to feed organic livestock (Greene et al. 2009). The unreliable supply of raw materials (such as certified organic grain needed for raising organic poultry and livestock) is one of the major factors limiting growth of the organic sector, especially organic meats (Dimitri and Oberholtzer 2009; Greene et al. 2009). The domestic supply shortage is occurring in tandem with increasing imports of organic foods, including off-season fruits and vegetables as well as other products like coffee and cocoa that do not grow in the continental United States (Liu 2011; Oberholtzer et al. 2012). Liu (2011) argues that the current regulatory framework is not only inadequate to the task of regulating domestic organics, but also incapable of ensuring the integrity of imported organics. Risks of low quality or falsely labeled goods appear to be particularly high when considering supplier nations, such as China, where corruption, environmental pollution, poor land tenure laws, and a lack of farmer empowerment characterize the agricultural systems (Liu 2011b).

Falling market prices for organic goods have contributed to fears that the price premium that underwrites organic production will fall too low to stimulate further conversion of farmlands to organic production, at a time when conversion is already too low to keep up with demand (Olson 2007). Furthermore, the diversification of alternative labels such as "natural," "cage-free," "hormone-free," GMO-free and others have created consumer confusion, ambivalence, and skepticism (Abrams et al. 2010; Van Loo et al. 2012; Schleenbecker and Hamm 2013). Finally, the diversification of the industry into processed and packaged foods that resemble "junk food" has raised consumer suspicions that corporations are undermining the quality of organic foods (Hartman Group 2014). Given this context, the viability of the market may be undermined by failures of the industry to provide products that meet consumer values and earn trust instead of generating doubts. Simultaneously, producers encounter pressures to lower prices to compete with cheaper "natural" products, which may threaten their livelihood as well as their ability to maintain the original values and goals of organic production (Kröger and Schäfer 2014).

Concluding Notes

The current expansion of US organic food markets continues a more-than-century-old trend in American society to question conventional agricultural production, which is associated with use of pesticides and other toxic chemicals, as well as soil erosion, environmental degradation, and economic disparities. Today, marketing for organic food recognizes that consumer decisions relate to a range of concerns that include individualistic interests in health, as well as altruistic concerns for environmental sustainability and social justice. Marketing strategies aim to expand sales by associating goods with consumers' expressed values. Even as organic markets continue to grow, increasing corporatization of certified organic goods and apparent greenwashing are creating doubts among consumers. Critics of organic markets point to contradictions between individualist and collective commitments for environmental and social well-being. In some cases, idealistic farmers and consumers are withdrawing from certified organic markets to seek approaches that would better align with their commitments to agroecology, social-environmental sustainability, and local food systems (Guthman 2014). In light of current trends and tensions, organic markets can be seen as a dynamic nexus in which corporate interests and consumers with diverse concerns struggle over the meaning, content, and future of organic foods.

References

Abrams, K. M., Meyers, C. A., and Irani, T. A. 2010. Naturally Confused: Consumers' Perceptions of All-Natural and Organic Pork Products. *Agriculture and Human Values* 27: 365–74.

Alkon, A. H. 2007. Growing Resistance: Food, Culture and the Mo' Better Foods Farmers' Market. *Gastronomica* 7: 93–9.

Alkon, A. H. and Norgaard, K. M. 2009. Breaking the Food Chains: An Investigation of Food Justice Activism. *Sociological Inquiry* 79: 289–305.

Aschemann, J., Hamm, U., Naspetti, S., and Zanoli, R. 2007. The Organic Market. In *Organic Farming: An International History*, ed. W. Lockeretz, 123–51. Oxfordshire, UK: CAB International.

Aschemann-Witzel, J., Maroscheck, N., and Hamm, U. 2013. Are Organic Consumers Preferring or Avoiding Foods with Nutrition and Health Claims? *Food Quality and Preference* 30: 68–76.

Atkinson, L., and Kim, Y. 2014. "I Drink It Anyway and I Know I Shouldn't": Understanding Green Consumers' Positive Evaluations of Norm-violating Non-green Products and Misleading Green Advertising. *Environmental Communication* 9: 37–57.

Belasco, W. J. 2007. *Appetite for Change: How the Counterculture Took on the Food Industry*. Ithaca: Cornell University Press.

Bellotti, E., and Panzone, L. 2016. Media Effects on Sustainable Food Consumption: How Newspaper Coverage Relates to Supermarket Expenditures. *International Journal of Consumer Studies* 40: 186–200.

Bellows, A. C., Onyango, B., Diamond, A., and Hallman, W. K. 2008. Understanding Consumer Interest in Organics: Production Values vs. Purchasing Behavior. *Journal of Agricultural and Food Industrial Organization* 6: 2–28.

Buck, D., Getz, C., and Guthman, J. 1997. From Farm to Table: The Organic Vegetable Commodity Chain of Northern California. *Sociologia Ruralis* 37: 3–20.

Clark, D. 2004. The Raw and the Rotten: Punk Cuisine. *Ethnology* 43: 19–31.

Curl, C. L., Beresford, S. A. A., Hajat, A., Kaufman, J. D., Moore, K., Nettleton, J. A., and Diez-Roux, A. V. 2013. Associations of Organic Produce Consumption with Socioeconomic Status and the Local Food Environment: Multi-Ethnic Study of Atherosclerosis (MESA). *PLoS ONE* 8: 1–8.

DiMatteo, K., and Gershuny, G. 2007. The Organic Trade Association. In *Organic Farming: An International History*, ed. W. Lockeretz, 253–63. Oxfordshire, UK: CAB International.

Dimitri, C., and Dettmann, R. L. 2012. Organic Food Consumers: What Do We Really Know about Them? *British Food Journal* 114: 1157–83.

Dimitri, C., and Lohr, L. 2007. The US Consumer Perspective on Organic Foods. In *Organic Food: Consumers' Choices and Farmers' Opportunities*, ed. M. Canavari and K. D. Olson, 157–67. New York: Springer.

Dimitri, C., and Oberholtzer, L. 2009. Marketing U.S. Organic Foods. www.ers.usda.gov/media/185272/eib58_1_.pdf

Fairtrade International. 2016. Monitoring the Scope and Benefits of Fairtrade, 7th edn. www.fairtrade.net/resources/monitoring-impact-reports.html (accessed October 23, 2016).

Fotopolous, C., A. Krystallis, and M. Ness. 2003. Wine Produced by Organic Grapes in Greece: Using Means-End Chains Analysis to Reveal Organic Buyers' Purchasing Motives in Comparison to Non-Buyers. *Food Quality and Preference* 14 (7), 549–566.

Friedland, M. T. 2005. You Call That Organic? The USDA's Misleading Food Regulations. *New York University Environmental Law Journal* 13: 379.

Greene, C. 2007. An Overview of Organic Agriculture in the United States. In *Organic Food: Consumers' Choices and Farmers' Opportunities*, ed. M. Canavari and K. D. Olson, 17–28. New York: Springer.

Greene, C., Dimitri, C., Lin, B.-H., McBride, W., Oberholtzer, L., and Smith, T. 2009. *Emerging Issues in the U.S. Organic Industry*. Washington, DC: US Department of Agriculture.

Guion, D. T., and Stanton, J. V. 2012. Advertising the U.S. National Organic Standard: A Well-Intentioned Cue Lost in the Shuffle? *Journal of Promotion Management* 18: 514–35.

Guptill, A. 2009. Exploring the Conventionalization of Organic Dairy: Trends and Counter-trends in Upstate New York. *Agriculture and Human Values* 26: 29–42.

Guthman, J. 2003. Fast Food/Organic Food: Reflexive Tastes and the Making of "Yuppie Chow." *Social and Cultural Geography* 4: 45–58.

Guthman, J. 2004. The Trouble with "Organic Lite" in California: A Rejoinder to the "Conventionalisation" Debate. *Sociologia Ruralis* 44: 301–16.

Guthman, J. 2008. Bringing Good Food to Others: Investigating the Subjects of Alternative Food Practice. *Cultural Geographies* 15: 431–47.

Guthman, J. 2014. *Agrarian Dreams: The Paradox of Organic Farming in California*, Vol. 2. Berkeley: University of California Press.

Hartman Group. 2010. Inspire Brand Love. www.hartman-group.com/hartbeat/348/inspire-brand-love (accessed October 19, 2016).

Hartman Group. 2014. As Organic's Authenticity Fades, Consumers Turn to Local Food. www.hartman-group.com/hartbeat/552/as-organic-s-authenticity-halo-fades-consumers-turn-to-local-food (accessed October 19, 2016).

Honkanen, P., Verplanken, B., and Olsen, S. O. 2006. Ethical Values and Motives Driving Organic Food Choice. *Journal of Consumer Behaviour* 5: 420–30.

Howard, P. H. 2009. Consolidation in the North American Organic Food Processing Sector, 1997 to 2007. *International Journal of Sociology of Agriculture and Food* 16: 13–30.

Howard, P. H. 2016. Organic Industry Structure: Acquisitions and Alliances. www.cornucopia.org/wp-content/uploads/2016/01/Organic-chart-Jan-2016.jpg (accessed January 2, 2017).

Hughner, R. S., McDonagh, P., Prothero, A., Shultz, C. J., and Stanton, J. 2007. Who Are Organic Food Consumers? A Compilation and Review of Why People Purchase Organic Food. *Journal of Consumer Behaviour* 6: 94–110.

Jaffee, D., and Howard, P. H. 2010. Corporate Cooptation of Organic and Fair Trade Standards. *Agriculture and Human Values* 27: 387–99.

Johnston, J., Biro, A., and MacKendrick, N. 2009. Lost in the Supermarket: The Corporate-Organic Foodscape and the Struggle for Food Democracy. *Antipode* 41: 509–32.

Kareklas, I., Carlson, J. R., and Muehling, D. D. 2014. "I Eat Organic for My Benefit and Yours": Egoistic and Altruistic Considerations for Purchasing Organic Food and Their Implications for Advertising Strategists. *Journal of Advertising* 43: 18–32.

Kröger, M., and Schäfer, M. 2014. Between Ideals and Reality: Development and Implementation of Fairness Standards in the Organic Food Sector. *Journal of Agricultural and Environmental Ethics* 27: 43–63.

Liu, C. L. 2011. Is "Organic" a Seal of Deceit? The Pitfalls of USDA Certified Organics Produced in the United States, China and Beyond. *Stanford Journal of International Law* 47: 333–78.

Lockeretz, W., ed. 2007. *Organic Farming: An International History*. Oxfordshire, UK: CAB International.

Lockie, S., Lyons, K., Lawrence, G., and Grice, J. 2004. Choosing Organics: A Path Analysis of Factors Underlying the Selection of Organic Food among Australian Consumers. *Appetite* 43: 135–46.

Lockie, S., Lyons, K., Lawrence, G., and Halpin, D. 2006. *Going Organic: Mobilizing Networks for Environmentally Responsible Food Production*. Wallingford: CAB International.

Makatouni, A. 2002. What Motivates Consumers to Buy Organic Food in the UK? *British Food Journal* 104: 345–52.

Martin, A. 2014. Wal-Mart Promises Organic Food for Everyone. Bloomberg.com. www.bloomberg.com/news/articles/2014-11-06/wal-mart-promises-organic-food-for-everyone (accessed April 4, 2016).

Mensing, B. 2008. USDA Organic: Ecopornography or a Label Worth Searching For? *Global Food & Agriculture* 9: 24,70.

Mintel. 2010. *Consumer Attitudes Toward Natural and Organic Food and Beverage – US – March 2010*. Mintel Group Market Research.

Mintel. 2011. *Natural and Organic Food and Beverage: The Market – US – October 2011*. Mintel Group Market Research.

Mintel. 2015. *Organic Food and Beverage Shoppers – US – March 2015*. Mintel Group Market Research.

Moberg, M., and Lyon, S. 2013. What's Fair? The Paradox of Seeking Justice through Markets.
    In *Fair Trade and Social Justice: Global Ethnographies*, ed. S. Lyon and M. Moberg, 1–23.
    New York: New York University Press.

Oberholtzer, L., Dimitri, C., and Jaenicke, E. C. 2012. International Trade of Organic Food: Evidence of
    US Imports. *Renewable Agriculture and Food Systems* 28: 255–62.

Olson, K. D. 2007. Current Issues in Organic Food: United States. In *Organic Food: Consumers'
    Choices and Farmers' Opportunities*, ed. M. Canavari and K. D. Olson, 185–93. New York: Springer.

Organic Trade Association. 2016. *State of the Organic Industry Fact Sheet 2016*. http://ota.com/sites/
    default/files/indexed_files/OTA_StateofIndustry_2016.pdf (accessed October 29, 2016).

Pearson, D., Henryks, J., and Jones, H. 2011. Organic Food: What We Know (and Do Not Know) about
    Consumers. *Renewable Agriculture and Food Systems* 26: 171–7.

Robinson, J. M. and Hartenfeld, J. A. 2007. *The Farmers' Market Book*. Bloomington: Indiana
    University Press.

Rodman, S. O., Palmer, A. M., Zachary, D. A., Hopkins, L. C., and Surkan, P. J. 2014. "They Just
    Say Organic Food Is Healthier": Perceptions of Healthy Food among Supermarket Shoppers in
    Southwest Baltimore. *Culture, Agriculture, Food and Environment* 36: 83–92.

Schleenbecker, R., and Hamm, U. 2013. Consumers' Perception of Organic Product Characteristics: A
    Review. *Appetite* 71: 420–29.

Schuldt, J. P., and Hannahan, M. 2013. When Good Deeds Leave a Bad Taste. Negative Inferences
    from Ethical Food Claims. *Appetite* 62: 76–83.

Sligh, M., and Cierpka, T. 2007. Organic Values. In *Organic Farming: An International History*, ed. W.
    Lockeretz, 30–39. Oxfordshire, UK: CAB International.

Sustainable Agriculture Research and Education. 2012. *Transitioning to Organic Production: History
    of Organic Farming in the United States*. www.sare.org/Learning-Center/Bulletins/Transitioning-
    to-Organic-Production/Text-Version/History-of-Organic-Farming-in-the-United-States (accessed
    October 23, 2016).

Tucker, C. M. 2011. *Coffee Culture: Local Experiences, Global Connections*. New York: Routledge.

USDA. 2016a. *Organic Agriculture: Organic Market Overview*. www.ers.usda.gov/topics/natural-
    resources-environment/organic-agriculture/organic-market-overview.aspx (accessed October
    23, 2016).

USDA. 2016b. *Farmers Markets and Direct-to-Consumer Marketing*. www.ams.usda.gov/services/
    local-regional/farmers-markets-and-direct-consumer-marketing (accessed January 2, 2017).

Van Loo, E. J., Caputo, V., Nayga, R., Canavari, M., and Ricke, S. C. 2012. Organic Meat Marketing. In
    Organic Meat Production and Processing, ed. S. C. Ricke, E. J. Van Loo, M. G. Johnson, and C. A.
    O'Bryan, 67–85. Oxford, UK: Wiley-Blackwell.

Van Loo, E. J., Caputo, V., Nayga, R. M., Meullenet, J.-F., and Ricke, S. C. 2011. Consumers'
    Willingness to Pay for Organic Chicken Breast: Evidence from Choice Experiment. *Food Quality
    and Preference* 22: 603–13.

Van Loo, E. J., Caputo, V., Nayga, R. M., Seo, H. S., Zhang B. Y., and Verbeke, W. 2015. Sustainability
    Labels on Coffee: Consumer Preferences, Willingness-to-Pay and Visual Attention to Attributes.
    Ecological Economics 118: 215–25.

Vogt, G. 2007. The Origins of Organic Agriculture. In *Organic Farming: An International History*, ed. W.
    Lockeretz, 9–29. Oxfordshire, UK: CAB International.

Wagner, K. 2015. Reading Packages: Social Semiotics on the Shelf. *Visual Communication*
    14: 193–220.

Wilk, R. 2010. Consumption Embedded in Culture and Language: Implications for Finding
    Sustainability. *Sustainability: Science Practice and Policy* 6: 38.

Zanoli, R., and Naspetti, S. 2002. Consumer Motivations in the Purchase of Organic Food. *British Food
    Journal* 104: 643–53.

15

Profile: Chef Steven Eckerd

Dr. Lang asks an important question: "Why might people want to pay more for organic and less for GMOs than conventionally grown foods?" This is a good question. Why do people choose to pay a premium for something, why do they value it more than another similar item? In restaurants, we ask ourselves this question all the time, because we are dependent on providing a meal—an experience—that people value and will pay money for. And by definition, a meal away from home, especially in fine dining, is always going to be more expensive than a home-cooked meal, and more expensive than many other food options like fast food or fast casual. Our customers choose to eat at our restaurant and chose to pay more for the food. What are the values that they bring to the table? What are the values that we offer?

To me the most important value we "bring to the table" is quality, which means using quality food products and treating them with creativity and respect, so that what is served to the diner is the best quality that we can provide. That's why our customers choose to spend more for a meal with us. But all they know is that the food tastes great, or looks good, and that the service is respectful. How do we communicate the quality of the product to them, so that they have a sense of how we value them, as well as the farmers and cooks who make the meal possible?

This chapter asks about quality and values and asks the reader to think through how they account for and value food, and how they value what they think they know even when what they know might be ambiguous. And while the issues of GMOs don't come up that much in my kitchens, the question of how you place a value on what isn't transparent is something we must think about when we cook for patrons. If people are worried about what's "in" their food if it's genetically modified, imagine how they must worry about what happens in a restaurant, when they can't see their food being made? The diner doesn't see the back of the house; they have to be able to trust us to serve them quality food even though they can't see us source it, prepare it, or plate it. They won't buy the meal if they don't trust what we do, so how do we encourage that trust? How do we signal to the diner that we are taking their health—and their pleasure in eating—seriously?

As Dr. Lang makes clear, some people assure themselves by buying organic since it can't contain GMOs per USDA law. But he also tells us that most whole foods (non-processed) don't have GMOs; so most of the food that a high-quality kitchen uses also won't contain GMOs (there are a few GMO vegetables: certain varieties of sweet corn and zucchini). Or, to be blunt, the meals

FIGURE 15.1 *Chef Steve with a freshly harvested squash. Photo courtesy of Janet Chrzan.*

produced by a restaurant that cooks from scratch won't contain GMOs. It's a dirty secret in the restaurant world that it's possible to order entire plates from the distributor; all the kitchen must do is microwave the food and present it as if it were their own product. And many restaurants do that, since it's cheaper than producing your own menu, in part because the kitchen staff doesn't have to be very well trained. But that's not what you should be finding in fine dining; we cook from scratch. And that's where my patrons have to trust me to source the very best food quality for them. For me, that means relying on local farmers for the bulk of our ingredients, but that requires a lot of coordination since farmers have opposite schedules from restaurants. And it means a lot of phone calls and texts to find out what's available and at peak so I can plan a menu.

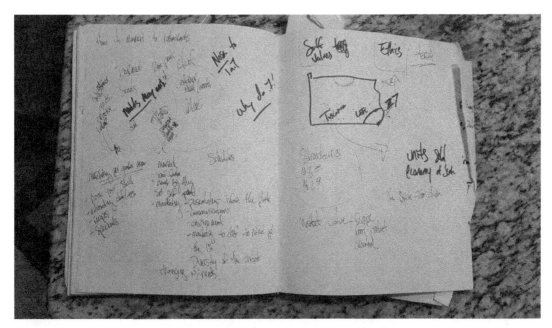

FIGURE 15.2 *Kitchen idea book. Photo courtesy of Steven Eckerd.*

Being anti-GMO is also currently trendy, as Dr. Lang describes. Trends are another issue in the restaurant business, since there are so many dining trends that patrons follow and if I'm producing food I have to know about those trends—what people want to eat—but also respect what's available and in season, my mission and vision, and my sources and staff. We walk a fine line, since customers follow trends on TV and media and expect to find that choice on the menu. But the pop culture trends last for roughly one to three years, and restaurant leases range from five to twenty years, so if you want to stay in business, you can't chase trends. You must present your own mission to attract customers, but still think about how health and diet trends affect the business. You must keep your vision but also create the WOW! factor, and you always compete with an ever and rapidly changing market. What's the next big thing and how do you adapt your product without changing your business? How do you adapt and stay creative while on mission?

I stay creative and ensure quality by buying from farmers I trust, whose product is excellent, in season, and represents the land. The flavors that fresh food provides allow me to be creative; you can see how I think food through by seeing the diagrams I construct in my idea book. But the most important element is the flavor of the food, the terroir, the passion and hard work the farmer puts into the fields, the dairy, or even the raising of animals, all of that ends up as the inspiration for the plate. When the food is sourced from someone I trust, my customers can trust me to provide the best for the land, for the plate, and for their health. Dr. Lang writes about the choices that need to be made when we eat; I think that when we are all producing our best with integrity we create trust in our food system.

FIGURE 15.3 *Chef Steve and farmer Alex Wenger with squash. Photo courtesy of Janet Chrzan.*

Is There Really a Difference between Conventional, Organic, and GMO?

John T. Lang

Suppose you wanted to buy some zucchini. Should you buy zucchini that is conventionally grown, organic, or genetically modified? Does it matter? Can you tell the difference between them? Why do they have different prices? Which should you choose? Why are GMOs so controversial and organic food is lauded? If I prefer one over the other, what questions should you ask? It seems that, regardless of their preferred diet, food shoppers are often confronted with a choice between conventional, organic, and GM food choices. But all is not as it seems.

Supermarkets are filled with all manner of processed food that contain at least some GM ingredients. In 1996, the first year of widespread commercialization, farmers in six countries planted more than one and a half million hectares of crops containing GM traits. By 2015, more than seventeen million farmers in twenty-eight countries planted more than 179 million hectares (roughly equivalent to the combined land mass of France, Japan, South Korea, Spain, and the United Kingdom) of crops containing GM traits (James 2015). Worldwide, almost 80 percent of all soybeans, a little less than three-quarters of all cotton, a little less than one-third of all maize (corn) and almost a quarter of all canola grown are genetically modified (James 2015). The five largest producers of crops with GM traits, accounting for 90 percent of the worldwide acreage, are the United States, Brazil, Argentina, India and Canada (James 2015). Including the twenty-eight countries that planted crops with GM traits, sixty-three countries in total have approved GM crops for food use, feed use, environmental release or planting (James 2015).

According to the US Department of Agriculture (USDA), GM varieties make up roughly 90 percent of all soybeans, more than 75 percent of cotton, and more than 80 percent of corn planted in the United States. Food manufacturers use these crops and their derivatives—such as high-fructose corn syrup, cottonseed oil, and soy lecithin—as ingredients in a vast array of processed foods. Moreover, GM sugar beets were first approved for planting in 2005, and now almost half of sugar used in the United States comes from this source. Consider the fact that the average supermarket stocks 30,000–40,000 food and beverage products (Nestle 2007). Assuming that three-fourths of processed food contains a GM ingredient, as the Pew Initiative on Food and Biotechnology estimated in 2005, then the average US supermarket stocks somewhere between 22,500 and 30,000 products that contain some GM ingredients.

While that might sound overwhelming, the vast majority of unprocessed foods contain no genetically engineered ingredients. In fact, the only GM foods that are grown and available to consumers are alfalfa, canola, corn (maize), cotton, papaya, potatoes, soybeans, sugar beets, and some types of squash. GM apples and GM salmon were approved in 2015 in the United States for human consumption, but were still not available for purchase in mid-2016.

Moreover, to meet the USDA's organic regulations, farmers and processors must show they are not using GM ingredients. That means that no certified organic products contain GM foods. If consumers wish to identify and avoid GM ingredients, they can buy 100 percent Organic, Certified Organic, and USDA Organic-labeled products. They can also look for the "non-GMO label" in their supermarket. This label comes from the Non-GMO Project, which is a private, nonprofit verification and labeling company. Though it can help consumers avoid GM products, this voluntary label has created a gray area, where products that currently do not have GM counterparts are being labeled as non-GMO. While technically accurate, should a product like popcorn be labeled as non-GMO, given that GM popcorn does not exist? Critics of this voluntary label argue that it implies that non-GM foods are better for you. Others point out that non-GMO agriculture is simply conventional agriculture that uses traditional agricultural pesticides and herbicides.

Conventional agriculture and farming are broadly used terms that generally refer to modern forms of agriculture that are not certified organic and use GM seeds. Exactly what form it takes depends on the individual farm, the region, and the country. In general, however, conventional farms use a variety of use of synthetic chemical pesticides and fertilizers to manage pests, weeds and fertility. Conventional agriculture is typically highly resource-intensive but also highly productive. To maximize efficiency, farmers will designate entire fields to just one crop, which creates uniformity. This uniformity reduces labor costs and makes harvesting easier. Over time, monocropping or duocropping, planting just one or two crops in a simple rotation on a particular piece of land, degrades soil health, can reduce biodiversity, and make crops more susceptible to ecological harm. But given its relative simplicity and economic advantages, modern farmers tend to specialize in one or two crops.

How Long Have We Been Genetically Modifying Food?

Some argue that Mendel's 1850s experiments, where he demonstrated that crossbreeding animals and plants could favor certain desirable traits, is a form of genetic modification. But the type of genetic modification that generates the most concern goes beyond the selective breeding techniques that farmers have used for centuries to modify the genes of organisms for crop improvements. More scientific, current versions of conventional hybridization often involve the use of DNA sequences associated with a particular gene or trait, complete with a known location on a chromosome, to track plants' genetic makeup during the development process. Generally, each successive hybridized generation has a mix of traits, both good and bad from their parents. Breeders try to build on the positive traits and minimize the negative traits over a number of breeding cycles.

Conventional plant breeding, even the modern variants, swaps many genes at once. As a result, breeders often discard thousands of plants, generation after generation, before finding one that results in a useful or noteworthy change. Breeders sometimes use mutagenesis of the breeding stock as a shortcut to this time intensive process. This technique hopes that something useful will result when seeds are exposed to chemicals or radiation that causes bits of DNA to copy incorrectly. The resulting mutations may have naturally occurred over time without this intervention, but mutation breeding speeds the process up and has led to commercial varieties of rice, wheat, and barley (Broad 2007), as well as Rio Red Grapefruit (Grun et al. 2004) Despite the fact that mutagenesis has changed an organism's DNA during the breeding process, it has not caused a popular uproar.

Selective breeding, marker-assisted hybridization, and mutagenesis are not what people are referring to when they colloquially say something is "genetically modified." In some ways, this shows the blurred line between biotechnology and non-biotechnology breeding. When people refer to GM crops, they are talking about a transgenic process where a scientist purposefully and directly manipulates a gene or DNA sequence using recombinant DNA technology. This might be a form of transgenesis, where scientists transplant genes between organisms that could not naturally breed or a form of cisgenesis, where scientists swap genes from the plant itself or a close relative, that could otherwise naturally breed. Regardless of the technique, seed scientists are generally trying to introduce new genetic characteristics to an organism to increase its usefulness.

Although there are several new genome engineering and synthetic biology tools and technologies, scientists have commonly used two methods to genetically modify plants. In the "gene gun" method, more properly known as the biolistic method, scientists shoot a metal pellet coated with DNA into a plant tissue target. The collision causes a few genes to be incorporated into the nucleus of the plant tissue. In another method, scientists also use a piece of DNA, called a plasmid, and then take advantage of the abilities of a naturally occurring soil microbe, generally *Agrobacterium tumefaciens* or *A. rhizogenes*, to transform plant cells to introduce new DNA into plant cells (Gelvin 2009; Barampuram and Zhang 2011).

Scientists have been manipulating and researching DNA sequences using relatively precise methods for more than thirty years. There are still parts of the process, however, that they do not fully understand and cannot fully control. Scientists do not control where the microbe inserts its DNA bundle; the bundle of DNA could fracture; or that unwanted genes could also be triggered. However, conventional plant breeding is also filled with considerable uncertainty. In reality, these same events can and do happen during conventional breeding. Moreover, to date, scientists have had less impact on a plant's genetic expression than conventional intervariety variation and environmental effects have had (Ricroch et al. 2011).

In contrast to these methods that have been typically used, newer techniques are now available to a greater number and greater variety of scientists. These newer options give additional control to scientists. They are fast, do not require as much specialized training, and are not as expensive. Clustered Regularly Interspaced Short Palindromic Repeats (CRISPR), in particular, has been credited moving the technology from a niche specialty to a mainstream method that can be used by a wide spectrum of biological researchers (Sander and Joung 2014). Although scientists noticed these short, repeating, palindromic DNA sequences separated by short, non-repeating "spacer" DNA sequences in 1987, it was not until 2012 that scientists demonstrated its genome-editing potential (Jinek et al. 2012). CRISPR can be used in seemingly any type of cell, not just plant cells, leaving experts concerned about the possibility of engineering human germ cells, including eggs, sperm, and embryos. Science and scientists are ill-equipped to evaluate these risks, uncertainties, and concerns alone. Groundbreaking research can spur all manner of ethical, legal, and social implications that sensibly give us reason to pause.

That is not to say that we no longer have to think about the risks and uncertainty with conventional breeding. But the more technical processes introduce some novel concerns. Conventional breeding and transgenic processes are different in two significant ways. First, scientists can isolate genes, and introduce new traits without simultaneously introducing many other traits, more easily using modern transgenic techniques. Second, scientists can cross biological boundaries that could not be crossed by traditional breeding, for example, transferring traits from bacteria or animals into plants.

Though scientists might want to help give crops increased resistance to things like extreme heat, drought, or soil nutrient deficiencies, to date they have modified GM crops to tolerate particular herbicides and resist specific pests. The best known example of insect resistance is the use of *Bacillus thuringiensis* (Bt) genes. Bt is a naturally occurring bacterium that produces crystal proteins lethal to insect larvae. Organic farmers and conventional farmers have used Bt as a natural pesticide for decades. By transferring Bt crystalline protein genes into corn plants, corn plants protect themselves by, in effect, producing their own pesticides. Farmers also need to protect their crops from weeds. In conventional farming, they spray herbicides before any of their crops or weeds emerge from the soil. With GM herbicide-tolerant crops, farmers can use spray their fields without fear of their herbicides killing their crops after they have emerged. This lets farmers spray the herbicide to kill weeds and not harm the plant. The most common herbicide-tolerant crops are resistant to glyphosate, commonly known by the trade name Round Up, an herbicide effective on many species of grasses, broadleaf weeds, and sedes.

Are They Regulated Differently? How Can Consumers Tell the Difference?

As a reaction to individual state laws about the labeling of GM food, in 2016, President Obama signed a federal law directing the USDA to create a national labeling standard for foods containing GMOs. Among other things, the USDA will decide what labels for GM ingredients should include; the options include one of the following: a few words on the label, a symbol that has not yet been designed, a toll-free telephone number for more information, or a QR code that can be scanned with a smartphone for more information. This law pre-empts state-labeling laws. Until the USDA completes its process, sometime in 2018, consumes can continue to rely on manufacturers' voluntary labels, retailers' programs and promises, or a certified organic label, to find out what is in the products they buy.

As part of the 1990 National Farm Bill, the USDA authorized the Organic Foods Production Act to create a national standard for organic farming practices and policies (National Organic Program 2000). In December 1997, the National Organic Program published the first proposed rules for national organic standards, which allowed the use of GM crops, sewage sludge, and irradiation in organic production. As a result, the USDA received an unprecedented outpouring of 280,000 mostly negative comments, and consequently the use of GM crops, sewage sludge, and irradiation were removed from the organic standards and are not allowable methods for organic production. An updated version of the proposed rule was published in the Federal Register on March 13, 2000, and after more public comments, a final rule that was published on December 21, 2000, with an effective starting date in April 2001, and full implementation by October 2002. Organic standards and rules are defined for the production process, rather than for the food. It takes a process-oriented approach since the composition of organic foods are not considered to be different than conventionally grown food. According to the final organic standard, organic farmers must forgo conventional synthetic pesticides and herbicides, petroleum-based and sewage sludge-based fertilizers, irradiation, genetic engineering (biotechnology), antibiotics, and growth hormones. Moreover, the land for certified organic farms must not have any prohibited substances applied to it for at least three years before harvesting an organic crop.

The rules for organic food are dynamic and undergo minor revisions on a regular basis. A citizens advisory group, the National Organics Standards Board, composed of members appointed by the Secretary of Agriculture for five-year terms, makes recommendations to the USDA about adding or removing items on the National List of Allowed and Prohibited Substances. This group must review every substance on the list every five years to make sure it meets the organic standards; they also consider petitions to add or delete particular substances. The resulting list defines the materials and farm inputs that are acceptable throughout the process, from farm to table. Individual states may also create their own organic programs, like the California Certified Organic Farmers program which was established prior to the national program. These state-level programs may be more restrictive than national requirements, though they must further the general purpose of the national program. State organic programs, and any future amendments to those programs, must be approved by the Secretary of the USDA before being implemented.

The public debate around the first proposed organic rule illustrates how contentious these debates have been, and how controversial the revisions continue to be. For example, there are currently fierce arguments about whether aquaponic and hydroponic growing methods that use containers but no soil should be eligible for organic status. Allowable since 2002, advocates argue that these methods enhance food safety and sustainability by using less water and less fertilizer; opponents argue that soil health is essential to organic farming and because these methods do not involve soil they should not be considered organic.

The regulations surrounding GM food predate organic food standards. The 1986 "Coordinated Framework for Regulation of Biotechnology" adapted existing food laws to GM food. Despite different breeding processes, the same regulations apply to genetically modified and conventional foods because the final products are considered to be substantially equivalent. Under this Coordinated Framework, the US Food and Drug Administration (FDA), the USDA, mainly through the Animal and Plant Health Inspection Service (APHIS), and the US Environmental Protection Agency (EPA) divided responsibility for evaluating and regulating GMOs. Depending on its characteristics, a product may be subject to review by one or more of these agencies. The FDA must evaluate the safety of new GM crop varieties before people or animals consume them. The USDA evaluates the potential impact of widespread release of the plant or crop, and monitors field trials. The EPA investigates both the human health and the environmental impacts of the pesticide levels in GM crops. Ultimately, as long as it does not introduce new allergens or substantially alter the nutritional value of the food, the United States treats GM foods as substantially equivalent to those produced conventionally. This means that the regulators are not trying to establish that GM food is absolutely safe. Rather, the regulatory agencies consider whether a GM food is as safe as its conventional, generally recognized as safe, counterpart.

The FDA's generally recognized as safe process allows manufactures to self-determine whether food ingredients are safe enough to be sold to consumers without prior government approval or restriction. This means, in effect, that many GM food ingredients can enter the food system through a streamlined process that does require any US government agency approval. This was made formal in a 1992 policy statement when the FDA clarified that it will treat GMO substances using "an approach identical in principle to that applied to foods developed by traditional plant breeding" (US FDA 1992). Notably, to help food manufacturers comply with their regulatory obligations, the FDA encourages them to participate in a voluntary premarket consultative process to address any potential issues.

Phillips McDougall, the agribusiness consulting company, estimated that the entire process of discovery, development, and authorization of a new plant biotechnology trait took anywhere from 11.7 to 16.3 years and cost $136 million (McDougall 2011). After identifying a desired trait, scientists attempt genetically isolate that trait, insert it into a target organism, and then propagate that organism. They also conduct preliminary toxicological and environmental tests hoping to yield a trait that could be commercially developed. Eventually, promising candidates are screened against a variety of pests, weeds, and diseases, in climate-controlled growth chambers, greenhouses, and laboratories. Along the way, companies use increasingly complex screening criteria to determine whether to continue research. Before moving to field trials, large or small, the developer submits a plan for minimizing cross-pollination to compatible plants that are not part of the field trial to Animal & Plant Health Inspection Service (APHIS). The companies that develop GM crops and seeds also perform extensive testing, comparing their composition to that of a non-modified, conventional version, before considering application for release. They submit a synopsis of these data, including an evaluation of genetic stability, compositional and nutritional analysis, as well as changes in allergenicity and toxicity, to regulators for review. Depending on the intended use of the organism, the FDA and the EPA may also assess the organism. Regulators also solicit public input and comments at several stages of the process. In practice this mean that, although there are some predictable steps, there is not a simple pathway that all firms or all GMOs follow as they move from creation to introduction.

The entire process continues to be convoluted in part because regulators were simply trying to adapt existing regulations to new technology when they conceived the original 1986 Coordinated Framework. Recognizing the intricacies that exist in the regulatory system, in 2015, the White House directed the EPA, FDA, and USDA to update the Coordinated Framework and develop a long-term strategy to make sure the regulatory system can handle future biotechnology developments (Holdren et al. 2015). In contrast, outside of the United States many other countries adhere to a precautionary principle, where new products or processes are resisted if their ultimate effects are disputed or unknown—in their regulations. In a process that can take more than two years, the European Union, for example, puts GMOs through a rigorous case-by-case, science-based food and environmental evaluation by the European Food Safety Authority (EFSA) before submitting a recommendation to any number of regulatory committees who can choose to accept or reject the EFSA proposal (Davison 2010). The recommendations are then subject to a number of political and legal decisions by member states. As a result, countries in the European Union have not, to date, cultivated significant quantities of GM crops; instead they mostly import and label GM foods. This does not make the issue any less complicated. For example, any product "produced from" a GMO must be labeled in the European Union, even if there is no detectable DNA or protein from the GMO in the final product, as is the case with soybean oil (European Union 2003). But, products like beer, wine, and cheese that may be derived through the use of genetically engineered yeast and do not have to be labeled (European Union 2003). This is just one example of a complex patchwork of divergent regulatory requirements. As such, GM foods are often subject to international political, legal, trade, and environmental disputes.

Why Don't They All Cost the Same? Why Might People Want to Pay More for Organic and Less for GMOs Than Conventionally Grown Foods?

Our modern food system, with its emphasis on mechanized, centralized, and globalized production has managed to produce enormous amounts and varieties of food available at our convenience at lower and lower prices to consumers. Food has dropped from more than 40 percent of an American's budget in 1900 to less than 20 percent in 1960 to less than 10 percent in 2015 (Thompson 2012; Barclay 2015). That said, what we choose to eat can affect our food costs.

Certified organic foods are generally more expensive than their conventional counterparts. Looking at a basket of one hundred common food items, one study found that organic foods and beverages cost an average 47 percent more in price than conventional alternatives (Cost of Organic Food 2015). This is true for a number of reasons. Though increasing, the organic food supply is relatively limited. In 2015, the United States had more than two million farms encompassing more than 900 million acres. As a subset, there are roughly 12,000 certified organic farms producing and selling $6.2 billion in organic commodities on 4.4 million acres. Because commodity crops account for the vast majority of land farmed, the overall share of organic farms remains very small. Even fruits and vegetables, which have a higher share devoted to organic production, are mostly produced through conventional agriculture.

Another reason that organic food tends to cost more than its conventional counterparts is because the production costs for organic food are generally higher. Increased human labor, managing weeds and monitoring pests, for example, create higher costs. This increased reliance of manual labor, along with a greater diversity of crops generally grown on organic farms, also means that economies of scale are difficult to achieve. Similarly, mandatory separation of conventional and organic crops and produce in processing and transportation results in higher costs because of the mandatory segregation of organic and conventional produce. Increasing the supply of organic food will slowly help reduce costs, though whether these savings are passed to consumers remains to be seen.

In turn, conventional foods are sometimes a bit more expensive than their GM counterparts. This is especially true if a food is labeled non-GMO; studies show that US consumers are willing to pay a small price premium for food certified to be free of genetic modification (Goodwin et al. 2015). But the reason for this is unclear and may simply reflect consumer's willingness to pay more to avoid GMOs than any production cost reasons. Food companies routinely reformulate their products without increasing retail prices. So, if companies reformulate products so that they do not contain GM ingredients, the cost of the food item does not necessarily need to change. For example, General Mills now makes Cheerios breakfast cereal without GM ingredients; they reformulated the cereal without increasing the price. This is because the prices that consumers pay at grocery stores are based on several factors beyond the costs. Shopper demographics, brand competition, and store characteristics, marketing costs, and so forth also dictate retail prices.

Why Are GMOs So Controversial and Organic Food Is Lauded?[1]

To some extent, GM food is controversial because critics have raised several environmental and human health safety concerns. They have, successfully in some cases, reframed genetic modification as a form of perverted science. Potent visual images, symbolic actions, and creative wordplay like the neologism "Frankenfood" has generated considerable media attention for those opposed to GMOs. Although long-term threats to human health and to the environment are difficult to assess, there is currently no firm evidence of serious environmental damage and no conclusive evidence of harm to human health. But more research should be done. Agribusiness has not generally supported, and in some cases even made it difficult, to conduct more research about these potential dangers. We can clearly say that the increased use of glyphosate (Roundup) has resulted in more glyphosate-resistant weeds; however, the intense use of glyphosate had led to less use of the other herbicides and, in some cases, slowing other forms of resistance.

Organic food is sometimes lauded because organic produce generally has lower levels of pesticide residue than conventional produce. Almost all produce, however, whether organic or conventional, has less pesticide residue than the maximum allowed by the EPA (Smith-Spangler et al. 2012). Similarly, despite slim evidence, the public believes that organic food is safer, more nutritious, and even better tasting that conventional food. The organic food industry has capitalized on these impressions in their marketing efforts, often romanticizing small farms and small-scale farming. As a result, labels can change people's perception of food. If a food was labeled as containing GMOs, people will be wary of it; if a food was labeled as organic, people would think it was healthier.

In both cases, moral entrepreneurs—anti-GMO activists on one hand and the organic industry on the other—have popularized public perceptions that help their own cause. At one level, the choice between genetically modified or organic food is about potential risks to human health and the environment. On a deeper level, however, the choices are about social and political power to shape our food system. Genetic modification is a marker for industrial, economically efficient agriculture that helps produce uniform and easily processed food. Organic food represents its antithesis. For those who oppose industrial agriculture or those who oppose genetic modification, organic agriculture presents a respectable alternative. This means that the debates between the systems can be thought of as a surrogate for the values that an individual or a society deems most important.

What Questions to Ask a Food Purveyor, a Farmer, a Food Producer?

We all depend on food with adequate nutrition and relatively free of harmful substances and pathogens. Our food system largely governs what foods we have available and their amounts and qualities. The history of food and agriculture—from conventional, genetically modified, or organic—in the United States illustrates such systems are neither accidental nor free of controversy. A wide variety of sociocultural, political, economic, and philosophical factors influence food production and consumption. Knowing what you value, and how you can support the systems that match your values, can help.

You might be inclined to just eat "natural" food. But keep in mind that "natural" food has no legal or agreed-upon definitions. Generally, you might expect that natural foods should have no artificial ingredients but there is no law to that effect. Orange juice provides an eye-opening example. If you wanted to make fresh squeezed orange juice, all you have to do is cut a few oranges in half and squeeze the juice—by hand, using a citrus reamer, or strainer if you prefer—into a glass. Depending on the variety of oranges, the time of year, and their quality, each glass you make might taste a bit different than the last one. It is easy to do, but most people prefer to buy orange juice. The steps are similar, but the industrial scale of production means that the tasks are a bit different. After washing, squeezing, and pasteurizing the oranges, food companies store the juice in million-gallon deaerated aseptic holding tanks (Jordán et al. 2003). Pasteurizing eliminates potentially harmful bacteria by heating the juice for a short period of time. Deaerating means that the oxygen is removed from the tanks so that the juice can be dispensed for up to a year without spoiling. But before packaging and shipping, each company adds their own proprietary "flavor pack" made from the chemicals that make up orange essence and oil. Not only do these packs ensure a distinct flavor profile for each brand of orange juice and give that brand a consistent taste month to month, it also allows to match the preferred taste of people in different countries (Hamilton 2010). These packs also compensate for the taste and aroma that are lost during the pasteurizing and storage process. This industrially processed and chemically enhanced orange juice is marketed as "pure," "simple" and made from "100 percent juice," but is it natural? Is it as natural as reasonably possible? Given the amount of food science involved in "natural" products, what we call natural is not as easy as it seems. Moreover, though it is often marked as healthy, being natural does not necessarily mean something is good for you.

Trade groups, lawyers, and government agencies have proposed solutions to the confusion. Some say we should try and define the words, some say we should sue people for using it in incorrect or misleading ways, while others argue that we should just ignore it. The reality of our food system is that people need to be informed about their food choices. The questions you ask as a consumer, and the values you choose to support with your money will become the future of our food. In general, agricultural and food production techniques mature and replace each other over decades. While a person's role as a consumer in a grocery store highlights individual, economic interactions, social and cultural expectations about safety, quality, taste, nutrition, as well as ethical aspects of food production and distribution are part of these interactions. Therefore, it is important to consider the supra-individual aspects of buying food to understand all elements that could be addressed for improving our food system, as opposed to simply supporting or rejecting conventionally grown, organically grown, or GM food.

At a deeper level, food touches on larger issues relating to social and political power, cultural values, corporate responsibility, and intellectual property. It is good to be informed about your food, and about the food system. But in reality, most people lack the knowledge, time, and inclination to figure it all out. Instead, they get guidance from people they trust, who they believe are honest, and whose values they share (Earle and Cvetkovich 1995). We might choose to listen to science or nutrition experts, or experts on our food system, but for many people, the most important experts might be religious, social, or ethical. People eat particular foods for a variety of reasons. For some, food is simply fuel or a collection of nutrients. But our food choices reflect our values as well as affect our health and our environment. As farmer and author Wendell Berry (1992: 374) fondly says, "eating is an agricultural act."

This means that finding a farmer or purveyor or set of food companies that match your values is paramount. Although such as sustainable agriculture and organic agriculture can be seen as competing paradigms to conventional agriculture or genetic modification, they would need to address the same problems of scale, geographical distribution, regulation, and market control as the conventional food system if they were to grow larger. For example, when Wal-Mart embraced organic products, food manufacturers like the Kellogg Company, PepsiCo Incorporated, and General Mills made plans to introduce organic products, necessitating large-scale production and distribution (Warner 2006). In the past we may have searched for one right way, but in the present and in the future, we will search for many right ways. You should find the choice that is right for you.

If you only eat certified organic food, you are avoiding GM ingredients. If you eat processed food, you are consuming GM ingredients. Does it matter which way you choose to eat? It seems like it should be easy to figure out, but this is not an easy task. What can scientists tell us? What can farmers tell us? What can regulators tell us? How is environmental harm measured? What are the long-term human health implications of your preferred form of agriculture? Are cultural values more important than economic considerations? What do *you* prioritize? Who do *you* trust to give us an honest evaluation of the pro-and-con arguments? Why? Does the development of GM foods and crops represent a prudent use of knowledge? What if it can help achieve humanitarian goals? What if its benefits and harms are unequally distributed? What if GM foods are neutral or beneficial for an individual but harmful to society or to the environment? Since organic farmers want to reduce pesticide use, and GM crops do allow farmers to use fewer, less harmful pesticides, should we reconsider using GM crops for organic certification? Sorting through these questions will give you a good start to think about the food system that reflects your values. Think of it as an opportunity to ask your farmer and your purveyor to improve them. We have created our food system through our choices, and we therefore have the power to change them.

Note

1 Material in this section and elsewhere draws on John T. Lang. 2016. *What's So Controversial about Genetically Modified Food?* London: Reaktion Books.

References

Barampuram, S., and Zhang, Z. J. 2011. Recent Advances in Plant Transformation. In *Plant Chromosome Engineering*, ed. J. A. Birchler, 1–35. New York: Humana Press.

Barclay, E. 2015. Your Grandparents Spent More of Their Money on Food Than You Do. www.npr.org/sections/thesalt/2015/03/02/389578089/your-grandparents-spent-more-of-their-money-on-food-than-you-do (accessed November 18, 2016).

Berry, W. 1992. The Pleasures of Eating. In *Cooking, Eating, Thinking: Transformative Philosophies of Food*, ed. D. W. Curtin and L. M. Heldke, 374–9. Bloomington: Indiana University Press.

Broad, W. J. 2007. Useful Mutants, Bred with Radiation. *New York Times*. www.nytimes.com/2007/08/28/science/28crop.html (accessed June 3, 2014).

Cost of Organic Food. 2015. *Consumer Reports*. www.consumerreports.org/cro/news/2015/03/cost-of-organic-food/index.htm (accessed November 18, 2016).

Davison, J. 2010. GM Plants: Science, Politics and EC Regulations. *Plant Science 178*: 94–98. https://doi.org/10.1016/j.plantsci.2009.12.005

Earle, T. C., and Cvetkovich, G. T. 1995. *Social Trust: Towards a Cosmopolitan Society*. Westport, CT: Praeger.

European Union. 2003. *Regulation (EC) No 1829/2003 of the European Parliament and of the Council of 22 September 2003 on Genetically Modified Food and Feed*. European Commission. October 18.

Gelvin, S. B. 2009. Agrobacterium in the Genomics Age. *Plant Physiology* 150: 1665–76. https://doi.org/10.1104/pp.109.139873

Goodwin, B. K., Marra, M. C., and Piggott, N. E. 2015. The Cost of a GMO-Free Market Basket of Food in the United States. *AgBioForum* 18: 25–33.

Grun, P., Ramsay, T., and Fedoroff, N. 2004. The Difficulties of Defining the Term "GM." *Science* 303: 1765–9. https://doi.org/10.1126/science.303.5665.1765b

Hamilton, A. 2010. *Squeezed: What You Don't Know about Orange Juice*. New Haven: Yale University Press.

Holdren, J. P., Shelanski, H., Vetter, D., and Goldfuss, C. 2015. Modernizing the Regulatory System for Biotechnology Products. www.whitehouse.gov/sites/default/files/microsites/ostp/modernizing_the_reg_system_for_biotech_products_memo_final.pdf

James, C. 2015. *Brief 51: 20th Anniversary (1996 to 2015) of the Global Commercialization of Biotech Crops and Biotech Crop Highlights in 2015*. Ithaca, NY: International Service for the Acquisition of Agri-biotech Applications.

Jinek, M., Chylinski, K., Fonfara, I., Hauer, M., Doudna, J. A., and Charpentier, E. 2012. A Programmable Dual-RNA–Guided DNA Endonuclease in Adaptive Bacterial Immunity. *Science* 337: 816–21. https://doi.org/10.1126/science.1225829

Jordán, M. J., Goodner, K. L., and Laencina, J. (2003). Deaeration and Pasteurization Effects on the Orange Juice Aromatic Fraction. *LWT—Food Science and Technology* 36: 391–6. https://doi.org/10.1016/S0023-6438(03)00041-0

Lang, J. T. 2016. *What's So Controversial about Genetically Modified Food?* London: Reaktion Books.

McDougall, P. 2011. *The Cost and Time Involved in the Discovery, Development and Authorisation of a New Plant Biotechnology Derived Trait: A Consultancy Study for Crop Life International*. http://croplife.org/wp-content/uploads/2014/04/Getting-a-Biotech-Crop-to-Market-Phillips-McDougall-Study.pdf

Nestle, M. 2007. *What to Eat*. New York: North Point Press.

Pew Initiative on Food and Biotechnology. 2005. *U.S. vs. EU: An Examination of the Trade Issues Surrounding Genetically Modified Food*. Washington, DC: Pew Charitable Trusts.

Ricroch, A. E., Bergé, J. B., and Kuntz, M. 2011. Evaluation of Genetically Engineered Crops Using Transcriptomic, Proteomic and Metabolomic Profiling Techniques. *Plant Physiology* 155: 1752–61. https://doi.org/10.1104/pp.111.173609

Sander, J. D., and Joung, J. K. 2014. CRISPR—Cas Systems for Editing, Regulating and Targeting Genomes. *Nature Biotechnology* 32: 347–55. https://doi.org/10.1038/nbt.2842

Smith-Spangler, C., Brandeau, M. L., Hunter, G. E., Bavinger, J. C., Pearson, M., Eschbach, P. J., … Bravata, D. M. 2012. Are Organic Foods Safer or Healthier Than Conventional Alternatives? A Systematic Review. *Annals of Internal Medicine* 157: 348–66. https://doi.org/10.7326/0003-4819-157-5-201209040-00007

Thompson, D. 2012. How America Spends Money: 100 Years in the Life of the Family Budget. *The Atlantic*. www.theatlantic.com/business/archive/2012/04/how-america-spends-money-100-years-in-the-life-of-the-family-budget/255475/ (accessed November 18, 2016).

USDA National Organic Program. 2000. 7 CFR 205.100 – 205.106–205.199. https://www.ecfr.gov/cgi-bin/text-idx?SID=cd8c7c24137bf150039c042e6ba263ef&mc=true&node=pt7.3.205&rgn=div5

US FDA. 1992. Statement of Policy—Foods Derived from New Plant Varieties: Guidance to Industry for Foods Derived from New Plant Varieties FDA Federal Register, Volume 57. www.fda.gov/Food/GuidanceRegulation/GuidanceDocumentsRegulatoryInformation/Biotechnology/ucm096095.htm (accessed December 23, 2016).

Warner, M. 2006. Wal-Mart Eyes Organic Foods. *New York Times*. www.nytimes.com/2006/05/12/business/12organic.html (accessed December 23, 2016).

16

Conclusion

E. N. Anderson, Jacqueline A. Ricotta, and Janet Chrzan

This volume began with a deep history of indigenous farming processes and ended with a discussion of genetically modified organisms. In between were chapters devoted to agronomy, history, philosophy, public policy, economics, marketing, and health. This was not an accident; clearly, if we are to understand organic food and farming we must start at the beginning and move forward, using the lens provided by multiple disciplines and many perspectives. Food production and utilization are, ultimately, made possible by the synergistic integration of biological and economic processes, but the forms, methods, and even conceptions of these processes are fully dictated and determined by human culture. Recognizing the importance of culture to human understanding of farming and food, this volume begins with a chapter by an anthropologist and ends with one written by a sociologist. In between are contributions by agronomists, economists, nutritionists, and practitioners, representing the myriad ways in which farming and food are linked and mutually deterministic. While this examination of organic food and farming is limited by the ability to cover such an integrally important human endeavor within the pages of even one book, let along hundreds, it is hoped that the chapters provide readers with sufficient overview to understand the depth and scope of the topic, and to begin to ask the comprehensive questions needed in order to continue the process of creating a sustainable, ecologically balanced, and economically fair agricultural system.

Agriculture is at a crossroads. Farmers are able to produce a surplus of food, but at what cost to the environment? And yet that surplus is not getting to the hungry. Gains in crop productivity continue to increase yields, but the changing climate has the potential to wipe out those gains. Use of synthetic fertilizers is at an all-time high, but nitrates toxic to human health can be found in drinking water across much of the corn belt. Sales of synthetic pesticides continue to climb, but so do rates of autoimmune and chronic diseases. However, what is also increasing, and at a rate that astounds food marketers, are the sales of organic food and fiber.

Is conventional agriculture, which feeds millions of people worldwide, sustainable? Or is a push towards a more organic system of production the savior that the organic pioneers like J. I. Rodale and Sir Albert Howard thought? By definition, all traditional intensive agriculture was sustainable.

If it were not, it would not have lasted long enough to be "traditional." However, sustainability is necessarily within limits. No system is sustainable if human population rises beyond a certain point. No system is sustainable if major shocks, such as massive climate change, intervene. Sustainability was often (perhaps always) learned the hard way. Many systems, when studied archaeologically, show a pattern of early wasteful and destructive use followed by a crash and a subsequent buildup of sustainable systems. Other systems simply grow better over time, as people develop new crops and better land use strategies. Conversely, people can change their systems in the direction of higher short-term yields at the expense of long-term stability. They frequently do this when times are hard, or when commercial opportunities become important. Regulation of use of natural resources in traditional societies is governed by cultural and religious rules and values, and proscriptions to inhibit overuse are part of the moral values necessary to support sustainability. People must, at the minimum, respect other lives, and develop a strong conscience, so that they do not take too much even when they "can get away with it." It appears from experience that the worst situation occurs when the land is owned by one party, but worked or used by another, with minimal local control or mutual discussion. The organic movement, as it has developed, is an attempt to take back the control of the land from large corporate entities to the smallholder farmer able to make a living, care for the environment and produce food from pieces of land formerly considered a "large yard" rather than a farm.

FIGURE 16.1 *Alex and Steve at the Field Edge Farm. Photo courtesy of Janet Chrzan.*

From the case studies presented in this book, and the thousands of other systems of intensive organic agriculture that traditional peoples have developed around the world, we can extract the principles of a valid organic agriculture. Many of these ideas are now being practiced in local development projects and are summarized in the Worldwatch Institute's *State of the World 2011*. In an earlier iteration of his excellent chapter in this volume, E. N. Anderson created this list to explain the characteristics of sustainable agricultural systems:

1. **Diversity**. The more diverse the system, the better it is buffered against pest and disease outbreaks, weather changes, and other stresses. Also, it can provide a broader base of benefits. Diversity is valuable not just overall, but in each field and plot. Diversity involves use of as many species as possible, but also the maintenance of a very large number of locally adapted varieties. A major cost of modernized agriculture has been loss of these "landraces," and even of specially developed varieties.

One of the greatest dangers to agriculture, over the long term, is epidemic disease. This propagates in monocultural stands, or even in small local plots if there is only one major crop grown. Worst is the situation in which only one *variety* of a species is grown, as was the case with potatoes in the Irish famine of 1846–8, and as is the case with bananas in much of the world today. In the future, with even more world travel and even more intensive agriculture, diversity will be the price of survival, and there is no sense pretending otherwise. We must not only eliminate monocropping but also go back to a world of small diverse fields. Organic farming, community farming, and, in general, any farming closely targeted for consumers can do far better than industrial farming at making multicropping and diversification profitable.

2. **Local adaptation**. Crop species and varieties (landraces) have been selected over the years—often over thousands of years—for local conditions. Local wild foods, medicinal herbs, and the like are known and are often brought into cultivation. For instance, wet rice agriculture failed in the Yucatan Peninsula because of the infertile limestone soils (as well as lack of local expertise). Maize-beans-squash-chile swiddening would not do well enough to succeed in wet rice country in Asia.

3. **Total use**. It is rare to find a plant species that traditional agronomists, such as the Maya, do not use. This contrasts dramatically with modern industrial agriculture, which usually focuses on one or a very few non-native crops, and often forces them to grow by creating an artificial, chemical-saturated environment. Also, traditional farmers use every part of an organism that they can. Consider the total use of the pig, from head to feet, that was the rule within living memory on American farms—and now is coming back in some restaurants.

4. **Total reuse**. Composting, efficient utilization, recycling, multiple uses for items, and general efficiency characterize all these traditional intensive systems.

5. **Fertilizing** is by any means possible. Overall, leguminous and other nitrogen-fixing plants, both tame and wild, are probably the major source of nitrogen in traditional (and organic) systems. The classic method, still widely used, is to rotate nitrogen-demanding crops with nitrogen-fixing ones, usually beans or alfalfa. Probably all intensive traditional cultivators use animal wastes, vegetable leftovers (often composted), and other by-products.

6. **Soil and water conservation**. Trees including orchards and woodlots, windbreaks, hedgerows, tree corridors, brush areas, herb gardens, any vegetation that significantly blocks the flow of water, all conserve water and soil substantially. So do "landesque capital" improvements like terracing (especially if stone-faced), leading water through carefully designed canal systems

(no straight shots from water head-on down), and lining irrigation canals. Making sure that draws, gullies, and the like are grown up to soil-holding brush is critically important. Organic material in the soil also holds soil and water. Intercropping and other intensive cropping systems can do this too. These methods are also encouraged by modern agroecology proponents.

7. **Pest control** is done by a number of means, many of them passive. Wild birds and predatory insects are left, and often encouraged or deliberately propagated. Native Americans set up birdhouses for martins (large swallows) that eat mosquitoes. The Chinese used to introduce predatory ants to citrus groves. Even today, blackberries are left growing around vineyards in California, because predatory insects breed in the blackberries and eat pests of vines. Much simpler and more widespread is breaking up monocultures or large blocks of cultivation. This can be done by crop rotation, small fields, burning over land before or after cropping, leaving many edges, having small fields with differing crops, multicropping, and doing everything else possible to maximize edges and variation over time and space; this disrupts pest cycles and prevents buildup of huge pest populations and of epidemics. The cleverest way to deal with pests is the south Chinese method: chickens or ducks are run through the fields, frogs are encouraged, catfish are tolerated at some stages of pond development—and then the pest control agents themselves become food.

8. **Full nutrition** from easy-to-grow, highly productive crops. Successful traditional agricultural systems are based on highly productive starch staples such as rice, wheat, or potatoes, with a backup such as maize, rye, or quinoa if the staple crop fails. They all have productive, successful, easily cultivated plant protein sources: soybeans in East Asia, common beans in the New World, many kinds of dal or pulse crops in India, etc. They also have animals. Critically important is having a range of vegetable, fruit, and leaf crops that are extremely high in vitamins and minerals, to prevent malnutrition on the usually starchy diet. Successful traditional systems have famine staples, including well-known wild food sources.

9. **Knowledge and skill**. These systems are also knowledge-intensive and skill-intensive. This helps with the labor problem: usually, the more skill, the less hand labor is needed. Knowledge of how to do things quickly and efficiently substitutes for endless repetitive inefficient handwork. The idea that traditional farmers are somehow ignorant or primitive could not be more wrong. It takes an exquisite knowledge of plants, animals, weather, seasons, soils, and other environmental factors to make a good living from harsh surroundings.

These observations written by an anthropologist eerily resemble the checklist of agroecological practices recommended by Worldwatch, UNEP-UNCTAD, and the FAO via the "Save and Grow" principles. Dr. Anderson is not a trained agronomist, but he is a trained observer specializing in food, agriculture, and economics. His suggestions for creating sustainability are rooted in successful cultural practices, while the recent agroecological recommendations rely on experimental and scientific agricultural practices with origins in traditional farming. Together they present three steps for improvement of current and future agriculture: observation of best practices, measurement of best practices, leading to improvements and even better practices. Furthermore, Anderson's analysis is stunningly like that of the agronomists Delate and Turnbull (this volume); all three reach very similar conclusions by asking very different questions, from completely different academic disciplines and epistemologies. Just as the earliest "modern" advocates of organics, such as Albert Howard and Gabrielle Matthei, William Albrecht, Lady Eve Balfour, Lord Northbourne, and Jerome Rodale drew their proposed practices from observation of successful indigenous farming methods, so must the agriculture of the future rely on established knowledges blended with tested new practices to ensure persistence and sustainability. Now, more than ever, we must learn from the past to create a blended workable future.

So what should the future of food production look like? Research from the Rodale Institute and Kathleen Delate and John Reganold point to the ability of organic farming systems, over the course of a few years, to match the yields of conventional grain crops such as corn and soybeans (for a review, see Chrzan in this volume). Another trial, the Marsden Farm Experiment (profiled by the Union of Concerned Scientists in May 2017; www.ucsusa.org/sites/default/files/attach/2017/05/rotating-crops-report-ucs-2017.pdf), points to a system of farming that, while not organic, seeks a hybridization of best practices that minimizes the environmental consequences of an industrial system while maintaining yields and productivity. In addition to small amounts of synthetic inputs, tools such as intensive crop rotation, use of cover crops, and integrated pest management are utilized to balance production and increase the sustainability of the farm as an ecological system. Above all, these systems salvage and protect crops by protecting and nourishing the soil. In effect, we are looping back to the analyses made by Robbins, admittedly a philosophical one, to argue that soil is not dirt but the matrix upon which all else rests. This point has been made by numerous agronomists, of course, but perhaps most cogently and concisely presented by John Reganold in his 2014 Ted talk (www.youtube.com/watch?v=NGBZZh8Oqyo). In it he highlights the need for integrated systems that creatively adopt best practices, and support for farmers who test and implement new methods to improve soil quality while increasing their farm's ecological health and maintaining farm profits.

Many examples of individual farms and research projects that conduct such inquiries exist, but perhaps one of the more interesting examples is the Tuscan estate Tenuta di Spannocchia (and not-for-profit organization) the Spannocchia Foundation. Both editors have visited and taught at Spannocchia, relying on the production methods developed there to teach about food and farming systems (www.spannocchia.org/). Spannocchia's mission is driven by a simple concept: "to create a sustainable future, you need to learn from the past." Spannocchia encourages global dialogue about sustaining cultural landscapes for future generations through "natural resource conservation, local ecology, sustainable agriculture and forestry, cultural history, traditional land management practices and farm based education." That translates into programming that blends the peasant farming techniques used for generations with modern practices designed to support the ecology of farming and food systems. The mixed-crop, certified organic farm models itself on the self-sustaining, inclusive farming model of historic Tuscany (the mezzadria system) but uses, where appropriate, the agricultural techniques of today, such as a high tunnel, integrated pest management, crop rotations, and rotational field and forest grazing for their landrace breed of swine, the Cinta Sinese, and Calvana beef cattle. Central to the Spannocchia model is living from the land as it was done hundreds of years ago. In addition to growing the majority of the food for the family, staff and visitors, the water for all plumbing is heated by fire from wood on the vast property; they have a twenty-year cutting plan. Even the building materials are sourced in situ, with stones from the fields, mortar made from local limestone, and clay quarried from the land to produce bricks and roof tiles.

It is also no accident that Spannocchia links cultural preservation to ecological preservation; the foundation recognizes that we must understand and include broader cultural ideals, values and norms in farming practice to ensure that the agricultural system makes cultural sense to

FIGURE 16.2 *The garden at Spannocchia. Photo courtesy of Janet Chrzan.*

FIGURE 16.3 *Weeding at Spannocchia. Photo courtesy of Jacqueline A. Ricotta.*

its users. Indeed, a quick reading of the goals of the organization reveals a list that includes some aspect of almost every topic covered in this volume, from agronomy, ecology, history, philosophy, economics, and marketing—all necessary to creating a farm and food system that is sustainable and supportive of the Triple Bottom Line. At Spannocchia, biological, economic, and human cultural systems are explicitly blended so that each may support the other, creating a farm and foundation synergy that is protecting both the history and the future of one very beautiful part of Italy.

But the viability and value of organic and integrated farming systems should not be assumed to exist only in protected areas of great beauty or in attractive, tourist-supported regions of traditional agricultural practice. On the contrary, hopefully this volume demonstrates that organic processes can and should penetrate every element of our food and farming systems, from scientific observation to agronomic method, to public, social, and health policy, to individual and community habits and values that protect the planet and ensure economic and social security of communities, cultures, and the land. That organic is not simply a dichotomized marketing tool but a deeply holistic, complicated and integrated means to understand how humans and the planet sustain and nourish each other. Hopefully this volume has provided readers with the intellectual wherewithal

FIGURE 16.4 *Produce from the Fields Edge Farm. Photo courtesy of Janet Chrzan.*

to sustain support of organic practices as well as to ask the social, cultural, and scientific questions necessary to create a more sustainable, supportive, secure, and biologically diverse agricultural economy.

Index